Lectures on Mathematical Combustion

J. D. BUCKMASTER
University of Illinois
and
G. S. S. LUDFORD
Cornell University

**SOCIETY for INDUSTRIAL and
APPLIED MATHEMATICS • 1983**

PHILADELPHIA, PENNSYLVANIA 19103

Copyright © 1983 by Society for Industrial and Applied Mathematics

Library of Congress Catalog Card Number: 83-61375

ISBN: 0-89871-186-X

Printed in Northern Ireland for the Society for Industrial and Applied Mathematics by The Universities Press (Belfast) Ltd.

Contents

Preface vii

Note on Notation viii

Lecture 1
PRE-ASYMPTOTIC COMBUSTION REVISITED
1. Ignition 1
2. Spontaneous combustion 2
3. Homogeneous explosion 4
4. Inhomogeneous explosion 6
5. Ignition by external agencies 10
6. Ignition by an externally generated hot spot 11

Lecture 2
GOVERNING EQUATIONS, ASYMPTOTICS, AND DEFLAGRATIONS
1. Equations for dilute mixtures 13
2. Nondimensional equations; Shvab–Zeldovich formulation 16
3. Activation-energy asymptotics 17
4. Plane deflagration waves 17
5. Generalizations 22

Lecture 3
GENERAL DEFLAGRATIONS
1. The hydrodynamic limit 23
2. Governing equations for the constant-density approximation 25
3. Slow variations with loss of heat 26
4. Multidimensional flames 30

Lecture 4
SVFs AND NEFs
1. Flame stretch 33
2. The basic equation for SVFs 34
3. The effect of stretch on SVFs 36
4. The basic equations for NEFs 38
5. NEFs near a stagnation point 40

Lecture 5
STABILITY OF THE PLANE DEFLAGRATION WAVE
1. Darrieus–Landau instability 45
2. The Lewis-number effect: SVFs 48
3. The Lewis-number effect: NEFs 50
4. The role of curvature 53

Lecture 6
CELLULAR FLAMES
1. Chaotic cellular structure 57
2. Effect of curvature 60
3. Flames near a stagnation point 62
4. Polyhedral flames 65
5. Other cellular flames 69

Lecture 7
PULSATING FLAMES
1. Solid combustion 71
2. The delta-function model 73
3. Stability of thermite flames 74
4. Flames anchored to burners 77
5. Stability of burner flames 79
6. Pulsations for rear stagnation-point flow 81

Lecture 8
COUNTERFLOW DIFFUSION FLAMES
1. Basic equations 85
2. The S-shaped burning response 87
3. General extinction analysis 90
4. Partial-burning branch 92
5. Stability 93
6. The ignition point 95

Lecture 9
SPHERICAL DIFFUSION FLAMES
1. Basic equations 97
2. Nearly adiabatic burning 100
3. General extinction and ignition analyses 102
4. Surface equilibrium 105

Lecture 10
FREE-BOUNDARY PROBLEMS
1. The hydrodynamic limit 109
2. The Burke–Schumann limit 110
3. NEF tips 112

4. NEF wall-quenching 115
5. Straining NEFs 119
6. Shearing NEFs 121
Appendix. The method of lines 122

References 125

Preface

The material contained herein is the written version of ten lectures on mathematical combustion given during a CBMS–NSF Regional Conference held at Colorado State University in June 1982. A few changes have been made, largely in the form of amplifications, but the result is faithful to what was delivered there. Even the lecture style is retained, which explains much and itself needs explanation.

We were, and are, primarily concerned with conveying the excitement of this new mathematical science. Full treatments of the material and pedantry over original sources (even though very recent) would not have contributed to such a goal. Most missing information can be found in our Cambridge monograph *Theory of Laminar Flames* (TLF), which appeared just after the conference. The remainder, corresponding to more recent topics, is referenced here. To the same end we limited ourselves to work that appeared to have unusual significance, studiously avoiding a review. If someone with a finger in the mathematical combustion pie is thereby offended, (s)he should know that we did it for the greater good.

The lectures were a five-day commercial for TLF; its preface should be read as a pre-preface here, giving the spirit of the enterprise. In no sense is the present text a reader's digest, however; if some phrases are the same (though we admit nothing), the reason is undoubtedly the uniqueness of perfection. Where the same topic is discussed, a fresh light has been thrown on it. In many instances, that was done automatically by adopting the so-called constant-density approximation from the start (and only abandoning it when absolutely necessary). But several topics here do not appear in TLF; examples are concentrated in the stability theory, which has surged forwards in the '80s. The theory was allocated one chapter of twelve in TLF; here it has commandeered thirty per cent of the lectures.

Jim Thomas and Aubrey Poore of Colorado State University should be thanked on two counts for arranging the Conference. Today's presses being what they are (and CUP did not strive to be an exception), a research monograph on a lively subject is sure to be dated by the time it reaches the bookshelves; such is the pace of modern research. The Conference gave us the opportunity to rewrite parts of TLF and update others. But it also enabled us to defend the TLF claim that combustion had finally reached the stature of a mathematical science, worthy of the attention of applied mathematicians. In that we were helped by other participants who gave complementary lectures; we would like to record their names here:

J. W. Bebernes	A. K. Kapila
J. F. Clarke	D. R. Kassoy
P. C. Fife	A. B. Poore

The preface would not be complete without mentioning the U.S. Army Research Office (ARO). The continued support of ARO-Mathematics, personified by Dr. Jagdish Chandra, is gratefully acknowledged.

December '82

J. D. Buckmaster
Urbana, Illinois

G. S. S. Ludford
Ithaca, New York

Note on notation. The notation has been kept as simple as possible. Rather than conscripting exotic symbols, we have made Latin and Greek letters do multiple duty; the meaning of a symbol should be clear from context. We have also been economical in the designation of and reference to lectures, sections, figures and displays (so as not to litter the page with numbers): except in figure captions, a decimal is used only for reference to another lecture. When part of a multiple display is intended, the letters a, b, ... are added to indicate the first, second, ... part.

To avoid repetition a uniform notation for asymptotic expansions in the small parameter θ^{-1} will be used throughout, namely

$$v = v_0 + \theta^{-1} v_1 + \theta^{-2} v_2 + \cdots$$

for the generic dependent variable v. The coefficients v_0, v_1, v_2, etc. will be called the leading term, (first) perturbation, second perturbation, etc. Only the variables on which these coefficients depend need be specified, changing as they do from region to region. In all cases the notation $+ \cdots$ is used (when a more accurate estimate is not needed) to denote a remainder that is of smaller order in θ than the preceding term. The same notation will be used for parameters; however, since their expansion coefficients are constants, no specification of variables is needed.

Similar conventions will be used when a small parameter other than θ^{-1} is involved or, in the case of $+ \cdots$, when asymptotic expansions in a variable are being considered. Except in §§ 2.1–2.3, numerical subscripts such as 0, 1, 2 will be used exclusively for terms in asymptotic expansions.

LECTURE 1

Pre-asymptotic Combustion Revisited

The description of reacting systems can be simplified when the so-called activation energy is large; the notion is an old one, but its full power is only realized by modern singular perturbation theory. More than forty years ago, Frank-Kamenetskii introduced approximations based on large activation energy to construct a thermal theory of spontaneous combustion, and we shall start there. His problem, which neglects the fluid-mechanical effects of main concern to us, focuses attention on the reaction and thereby acts as a precursor for the lectures that follow. The problem and its generalizations have been the happy hunting grounds of mathematical analysts for many years, but it was not until quite recently that a complete description of the ignition and explosion processes was made available by Kapila and Kassoy (working separately) through activation-energy asymptotics, the main theme of these lectures.

1. Ignition. Let us suppose that a combustion system has a characteristic temperature T_c and that the heat generated by reaction can be expressed as a function of T_c in the Arrhenius form

$$qe^{-\theta/T_c}. \tag{1}$$

This function has an inflection point at $T_c = \theta/2$ and its second derivative is positive for smaller values, where the graph is accordingly concave upwards (Fig. 1). Suppose also that the heat loss by conduction and convection has the linear form

$$k(T_c - T_f) \tag{2}$$

where T_f is the ambient temperature.

The system can only be in equilibrium if the heat generated (1) is equal to the heat lost (2). The parameters q and k do not then play independent roles, but rather, it is

$$\mathscr{D} = \frac{q}{k} \tag{3}$$

that is relevant. This ratio, which will be called the Damköhler number (cf. later lectures), can be altered by changing the parameters of the system.

It is apparent from Fig. 1 that, for $\theta > 4T_f$, either there is one solution or there are three solutions, depending on the value of $\mathscr{D}$. If $\mathscr{D}$ is increased, the straight line rotates about its end point in a clockwise direction, and we can identify the transitions from 1 solution (cold), to 3 solutions, to 1 solution (hot). It is this second transition that is our concern in this lecture. The state of the system, represented by a cold point such as C, moves towards I as the

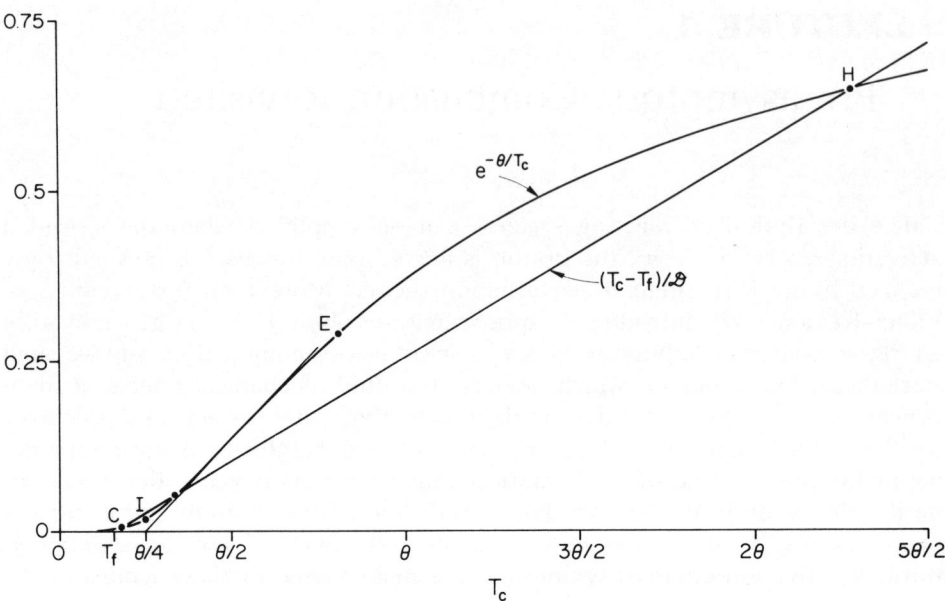

FIG. 1.1. *Heat-generation and heat-loss curves.*

Damköhler number is increased, and then must jump to a hot point such as H at the transition. This jump is called ignition.

Ignition is ubiquitous in combustion systems; it can generally be attributed to the nonlinear dependence of heat generation on temperature and the essentially linear dependence of heat loss. The precise nature of the phenomenon can only be determined by detailed analysis, though the results of different calculations carried out by activation-energy asymptotics often bear a strong family resemblance. They are characterized by the following elementary example, introduced by Frank-Kamenetskii.

2. Spontaneous combustion. Consider the boundary-value problem

$$\frac{d^2T}{dx^2} = -\mathcal{D}e^{-\theta/T} \quad \text{for } |x| \leq 1, \qquad T = T_f \quad \text{at } x = \pm 1. \tag{4}$$

Heat conduction in the infinite slab is balanced by heat generation due to the reaction. Depletion of the reactant has been ignored, so that the reaction rate depends only on temperature, as in § 1.

Such models have been used for many years to explain spontaneous combustion, the auto-ignition that occurs, for example, in large volumes of damp organic material. Nondimensionalization of the heat-conduction Laplacian makes the Damköhler number $\mathcal{D}$ proportional to a^2, where a is a length characteristic of the volume (here the semithickness of the slab). Thus, an increase in $\mathscr{D}$ may be achieved by increasing the volume and, as we shall see, this can lead to ignition.

We seek a solution of the problem (4) that, as $\theta \to \infty$, deviates only by $O(\theta^{-1})$ from the uniform state, i.e.

$$T = T_f + \theta^{-1} T_f^2 \phi + \cdots \quad \text{with } \phi = -\left(\frac{1}{T}\right)_1. \tag{5}$$

This leads to

$$\frac{d^2\phi}{dx^2} = -\delta e^\phi \quad \text{with } \delta = \frac{\mathscr{D}\theta e^{-\phi/T_f}}{T_f^2}, \tag{6}$$

an equation first obtained by Frank-Kamenetskii, and the boundary conditions

$$\phi = 0 \quad \text{at } x = \pm 1. \tag{7}$$

Here δ, the scaled Damköhler number, is assumed to be $O(1)$.

The perturbation ϕ achieves its maximum (ϕ_m) at the midpoint $x = 0$ and so may be written

$$\phi = 2\ln[e^{\phi_m/2} \operatorname{sech}(cx)] \quad \text{with } c^2 = \tfrac{1}{2}\delta e^{\phi_m}. \tag{8}$$

The boundary conditions then imply that

$$\sqrt{\frac{\delta}{2}} = e^{-\phi_m/2} \cosh^{-1}(e^{\phi_m/2}), \tag{9}$$

which defines the maximum temperature in terms of the parameter δ. The result is shown in Fig. 2, displaying the phenomenon of ignition. For δ less than the critical value

$$\delta_c = 0.878, \tag{10}$$

there is a steady solution, in fact there are two solutions, whereas for $\delta > \delta_c$ there is no solution of the type (5), and in fact none at all. The absence of a steady state for supercritical values of δ implies that, with unsteady effects included, the temperature will increase without bound when, for example, the system is initially in a uniform state $T = T_f$. In practice, the increase is limited by depletion of the reactant, an effect that is ignored here. Of the two solution branches for $\delta < \delta_c$, the upper one is believed to be unstable, though this has never been proved.

It is a general characteristic of ignition that it is associated with $O(\theta^{-1})$ perturbations of the frozen solution, i.e. the solution obtained for $\mathscr{D} = 0$. This is certainly true for the diffusion flames treated in Lectures 8 and 9 (see § 8.6); the details differ from those presented here, but the essential ideas do not.

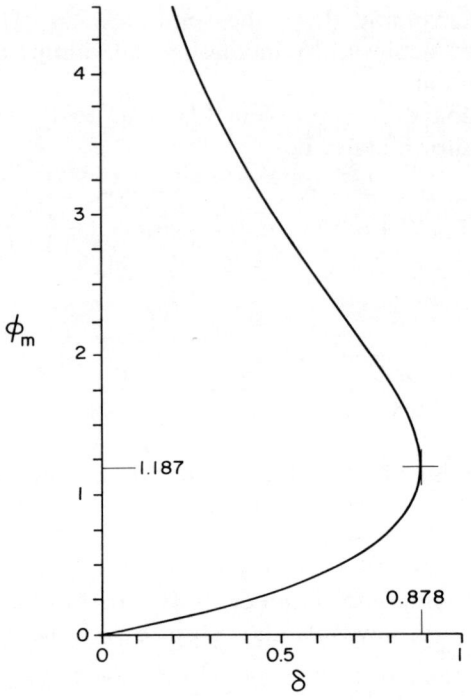

Fig. 1.2. *Steady-state response for slab with surfaces maintained at initial uniform temperature, as determined by* (9).

3. Homogeneous explosion. So far we have inferred ignition from a steady-state theory. The phenomenon itself is inherently unsteady, and certain aspects of the unsteadiness deserve examination. To that end, it is useful to consider first the spatially homogeneous initial value problem

$$\frac{dT}{dt} = \mathcal{D}e^{-\theta/T}, \qquad T = T_f \quad \text{for } t = 0. \tag{11}$$

There is no value of $\mathcal{D}$ for which a steady state can be attained; the problem is always supercritical. The physical reason is that no heat-loss mechanism, such as conduction to the boundaries (§ 2), exists.

In terms of the exponential integral

$$\text{Ei}(y) = \rlap{\,/}{\int_{-\infty}^{y}} \frac{e^u}{u} \, du, \tag{12}$$

this has the exact solution

$$\mathcal{D}t = \theta[f(T_f) - f(T)] \quad \text{with } f(y) = \text{Ei}\left(\frac{\theta}{y}\right) - \frac{y}{\theta} e^{\theta/y} \tag{13}$$

for any value of θ. Since the function f has the asymptotic expansion

$$f(y) = \left(\frac{y}{\theta}\right)^2 e^{\theta/y} + \cdots \quad \text{as } \theta \to \infty, \tag{14}$$

it follows that

$$\tilde{t} = \tilde{t}_e - \left(\frac{T}{T_f}\right)^2 e^{\theta/T - \theta/T_f} + \cdots \quad \text{with } \tilde{t} = \delta t; \tag{15}$$

here

$$\tilde{t}_e = 1. \tag{16}$$

Deviations from the initial state of the form (5) are therefore described by

$$\phi = -\ln(\tilde{t}_e - \tilde{t}), \tag{17}$$

so that t_e is the time to explosion (i.e. the time that T takes to deviate from its initial value T_f by more than $O(\theta^{-1})$). The behavior of T in some small neighborhood of t_e is called thermal runaway.

Thermal runaway terminates what is known as the induction phase, the problem with which pre-asymptotic theory was almost exclusively concerned. To go substantially further, modern asymptotic methods must be used, as will be discussed in these lectures. To analyze the so-called explosion phase that follows induction, we return to the expansion (15), and introduce a fast time τ given by

$$e^{-\theta\tau} = \tilde{t}_e - \tilde{t}. \tag{18}$$

This identifies an exponentially small neighborhood of t_e within which the expansion can be written in the form

$$\frac{1}{T} - \frac{1}{T_f} = -\tau + 2\theta^{-1} \ln\left(\frac{T_f}{T}\right) + \cdots, \tag{19}$$

so that, to leading order,

$$T = \frac{T_f}{1 - \tau T_f}. \tag{20}$$

Starting at the value T_f for $\tau = 0$, the temperature increases without bound as τ increases to $1/T_f$. The unboundedness is a consequence of our failure to account for reactant depletion; if that is remedied, T increases towards a burnt value T_b (see (25)), entering the so-called relaxation phase when it is $O(\theta^{-1})$ away. The relaxation phase lasts an exponentially short time also.

These features are shown in Fig. 3 and, so long as $T - T_b = O(1)$, the problem without depletion provides a qualitatively accurate description of them. In particular, the fast time τ is still relevant.

The results can also be obtained directly, without recourse to the exact solution. Thus, if the expansion (5a) is substituted into the problem (11), we

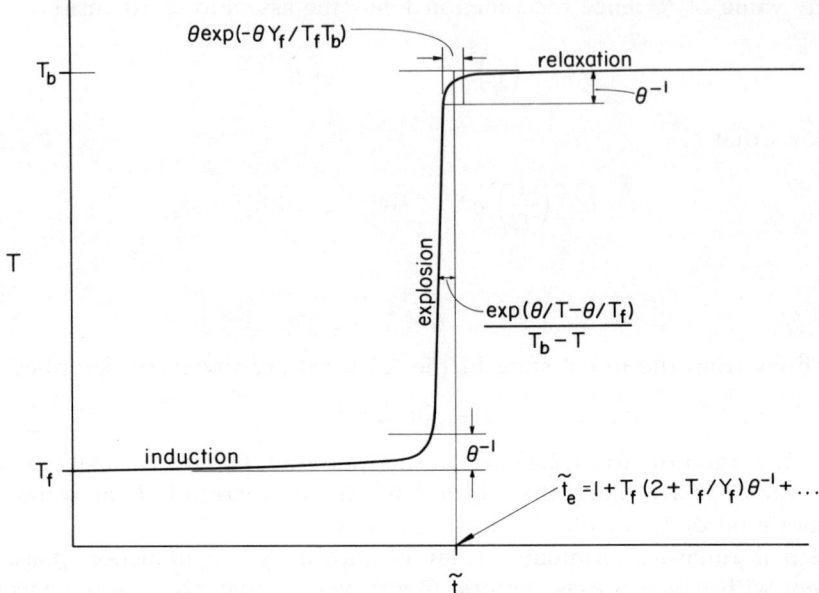

Fig. 1.3. *Temperature history for homogeneous explosion with reactant depletion.*

find

$$\frac{d\phi}{d\tilde{t}} = e^{\phi}, \qquad \phi = 0 \quad \text{for } \tilde{t} = 0, \tag{21}$$

with solution (17); and introducing the fast time τ into equation (11a) yields

$$\frac{dT}{d\tau} = T_f^2 e^{\theta(1/T_f - 1/T - \tau)}, \tag{22}$$

with solution (19).

4. Inhomogeneous explosion. We now combine unsteadiness with spatial inhomogeneity by considering the slab problem

$$\frac{\partial T}{\partial t} - \frac{\partial^2 T}{\partial x^2} = \frac{T_f^2 \delta}{Y_f \theta}(T_b - T) e^{\theta/T_f - \theta/T}, \tag{23}$$

$$T = T_f \quad \text{for } t = 0 \text{ and } x = \pm 1. \tag{24}$$

To account for reactant depletion, $\mathscr{D}$ has been replaced by $\mathscr{D}Y/Y_f$, where Y is the mass fraction (i.e. concentration) of the reactant and Y_f its initial value. The Shvab–Zeldovich relation

$$T + Y = T_f + Y_f \equiv T_b \tag{25}$$

(cf. § 2.2) is then used to eliminate Y, thereby ensuring conservation of total enthalpy (the sum of thermal enthalpy, represented by T, and chemical enthalpy, represented by Y).

During the induction phase, the $O(\theta^{-1})$ departures of T from T_f expressed by the expansion (5) satisfy

$$\frac{\partial \phi}{\partial t} - \frac{\partial^2 \phi}{\partial x^2} = \delta e^\phi, \qquad \phi = 0 \quad \text{for } t = 0 \text{ and } x = \pm 1. \tag{26}$$

Reactant depletion plays no role during this initial evolution of the temperature. When the system is subcritical, i.e. for $\delta < \delta_c$, the perturbation ϕ tends to the steady state (8) with the smaller ϕ_m, as $t \to \infty$. But, for $\delta > \delta_c$, the absence of a steady-state solution implies that ϕ will increase without limit, and indeed thermal runaway is found (numerically) to occur after a finite time.

Further progress depends on a description of this runaway, which by symmetry must take place in the neighborhood of $x = 0$. Since the spatial derivatives must play a role, they have to be increasingly large in order to be comparable to the ever-increasing time derivative. It follows that the region in which runaway occurs, called a hot spot, must continually shrink; this self-focusing is an essential feature of the process.

The appropriate variables for the runaway are $\tilde{t}$ and

$$\eta = \frac{x}{(t_e - t)^{1/2}} \tag{27}$$

(see Fig. 4), where $t_e(\delta)$ is the runaway time, to be determined numerically. The use of η is suggested by the form of (26) and by the focusing discussed above, though Kassoy's achievement in identifying it is in no way diminished

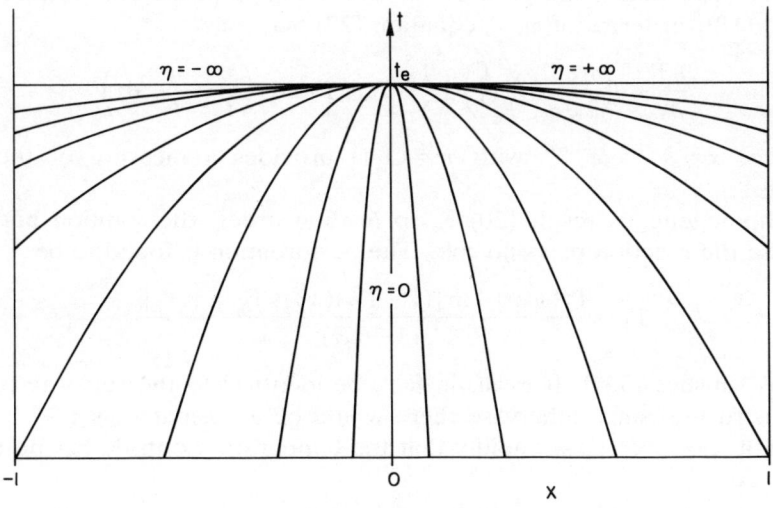

FIG. 1.4. Parabolas $\eta = \text{const}$.

by such a posteriori observations. Now we have

$$\frac{\partial \phi}{\partial \tilde{t}} - (\tilde{t}_e - \tilde{t})^{-1}\left[\frac{\partial^2 \phi}{\partial \eta^2} - \frac{\eta}{2}\frac{\partial \phi}{\partial \eta}\right] = e^\phi \qquad (28)$$

and we seek an asymptotic expansion

$$\phi = -\ln(\tilde{t}_e - \tilde{t}) + \psi(\eta) + \cdots \quad \text{as } \tilde{t} \uparrow \tilde{t}_e \text{ with } \eta \text{ fixed,} \qquad (29)$$

finding

$$\psi'' - \tfrac{1}{2}\eta\psi' + e^\psi = 1 \quad \text{with } \psi'(0) = 0 \qquad (30)$$

(as a symmetry condition). Another boundary condition is needed to complete the problem for ψ, and this comes from matching with the solution outside the shrinking hot spot, i.e. as $\eta \to \infty$ with x fixed. Thus ψ has the asymptotic form

$$\psi = -2\ln\eta + A + \cdots \quad \text{as } \eta \to +\infty, \qquad (31)$$

corresponding to

$$\phi = -2\ln x - \ln\delta + A + \cdots \quad \text{as } \tilde{t} \uparrow \tilde{t}_e \text{ with } x \text{ fixed.} \qquad (32)$$

Numerical solutions of the supercritical problem (26) exhibit the behavior (32), and hence determine the constant $A(\delta)$ that is needed for the second boundary condition

$$\lim_{\eta \to \infty}(\psi + 2\ln\eta) = A(\delta). \qquad (33)$$

The problem (30), (33) uniquely determines the function ψ, which can be readily calculated numerically.

Following the initiation of thermal runaway, the temperature rises at an increasing (exponential) rate, so that self-focusing continues. The variable η still plays a role during this process but now the appropriate time variable is the fast one (18); in terms of η, τ, equation (23) becomes

$$\frac{\partial T}{\partial \tau} + \theta\left(\frac{\eta}{2}\frac{\partial T}{\partial \eta} - \frac{\partial^2 T}{\partial \eta^2}\right) = \frac{T_f^2}{Y_f}(T_b - T)\exp\left(\frac{\theta}{T_f} - \frac{\theta}{T} - \theta\tau\right). \qquad (34)$$

Note that $x = \delta^{-1/2}\eta e^{-\theta\tau/2}$ with $\eta = O(1)$ provides a measure of the rapid focusing.

The homogeneous result (20) is, to leading order, the solution here also; otherwise the reaction plays no role. The perturbation is found to be

$$T_1 = \frac{T_f^2\{\psi(\eta) - \ln[(1 - T_f\tau)(Y_f - T_b T_f\tau)/Y_f]\}}{(1 - T_f\tau)^2}, \qquad (35)$$

where ψ satisfies (30a). It must, in fact, be identical to the runaway function just constructed, since otherwise there would be a mismatch as $\tau \to 0$.

The hot spot evolves so rapidly that the temperature outside has no time to change, i.e.

$$T = T_f + \theta^{-1}T_f^2(-2\ln x - \ln\delta + A) + \cdots \quad \text{as } x \to 0, \qquad (36)$$

and this does not match the hot-spot expansion, even to leading order. The reason is clear: the expansion (36) breaks down at points inside the hot spot when it is thickest (τ small), and such points can be well outside once the focusing is under way (τ moderate). The shrinking hot spot leaves behind an intermediate structure, which turns out to be stationary (i.e. independent of τ); to leading order it is described by

$$T = \frac{T_f}{(1 - T_f \chi)} \quad \text{with } \chi = -2\theta^{-1} \ln x. \tag{37}$$

For the homogeneous problem of §2, with no reactant depletion, the temperature increases indefinitely as $\tau \to 1/T_f$; but, with depletion, Fig. 3 shows that T is limited by the value T_b, which is approached within $O(\theta^{-1})$ as $\tau \to Y_f/T_b T_f$ (an earlier time). The latter is true here also, but there can no longer be the single relaxation phase shown in Fig. 3 as being described on the scale

$$\bar{\tau} = \frac{(\tilde{t} - \tilde{t}_e)}{\varepsilon} \quad \text{with } \varepsilon = \frac{Y_f \theta}{T_f^2} e^{-\theta Y_f / T_f T_b}, \tag{38}$$

since now the temperature is close to T_b only in the vicinity of $x = 0$. Instead, there is a transition phase at the hot spot, described in terms of $\bar{\tau}$ and the spatial variable

$$\bar{\chi} = \frac{\delta^{1/2} x}{\varepsilon^{1/2}}, \tag{39}$$

followed by propagation and, finally, relaxation phases. Figure 5 is the result of numerically integrating the limit problem for the transition phase, during which the focusing of the hot spot is opposed by reactant depletion, thereby forming an incipient deflagration wave (§ 2.4). Once formed, the wave propagates rapidly through the right side of the slab, burning up the reactant, after which the relaxation phase takes over. (Of course, similar remarks apply to the

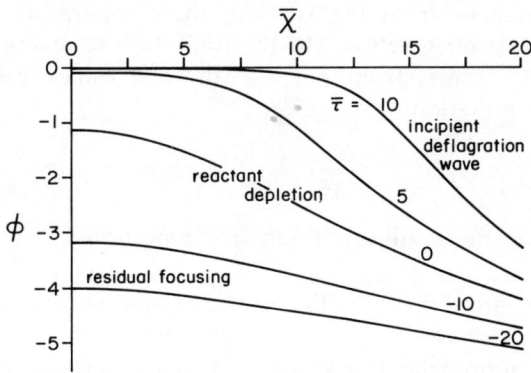

FIG. 1.5. *Temperature profiles during transition phase.* (Courtesy A. K. Kapila.)

left side of the slab too.) Buckmaster & Ludford (1982, p. 236) give the details, following Kapila.

5. Ignition by external agencies. So far we have confined the discussion to ignition due to self-heating, but it can also be caused by an external energy. As an example, suppose the half-space $x<0$ is filled with a combustible material subject to a prescribed heat flux at its surface. The mathematical problem is

$$\frac{\partial T}{\partial t} - \frac{\partial^2 T}{\partial x^2} = \mathcal{D} e^{-\theta/T}, \tag{40}$$

$$\frac{\partial T}{\partial x} = T'_s > 0 \quad \text{at } x=0, \qquad T = T_f \quad \text{for } t=0 \text{ and as } x \to -\infty. \tag{41}$$

Here the constant T'_s is the dimensionless heat flux and we shall suppose that

$$\mathcal{D} = e^{\theta/T_r} \quad \text{with } T_r > T_f. \tag{42}$$

The parameter T_r characterizes the reactivity of the material: for temperatures below T_r, the reaction is negligibly weak in the limit $\theta \to \infty$. A pre-exponential factor (even depending on θ) can be given to $\mathcal{D}$, but this is equivalent to changing T_r slightly, provided nothing is added to the exponential growth of $\mathcal{D}$ with θ.

During an initial (finite) time interval, the material is colder than the reactivity temperature so that the reaction is frozen (i.e. exponentially weak) and the heat equation governs. Because of the heat flux T'_s the temperature rises, its maximum value occurring at the surface $x=0$. Ignition occurs when the surface temperature reaches T_r. The subsequent process of thermal runaway, hot-spot development and deflagration-wave formation has been discussed by Kapila. Here we shall mention only the mathematical problem involved in the thermal runaway.

The rise in temperature is much more rapid than that for spontaneous combustion, extending only over a period $O(\theta^{-1})$. If conduction is to rival the temporal changes, spatial gradients must therefore be $O(\theta^{1/2})$ so that, at corresponding distances from the surface, the temperature has dropped by $O(\theta^{-1/2})$ and there is no reaction. (In the limit $\theta \to \infty$, reaction only occurs at temperatures $O(\theta^{-1})$ away from T_r.) The thermal runaway is therefore governed by the heat equation

$$\frac{\partial \tilde{T}_2}{\partial \tilde{\tau}} - \frac{\partial^2 \tilde{T}_2}{\partial \tilde{\chi}^2} = 0 \tag{43}$$

and its cause lies in the nonlinear boundary conditions

$$\frac{\partial \tilde{T}_2}{\partial \tilde{\chi}} = e^{\tilde{\tau} + \tilde{T}_2} \quad \text{for } \tilde{\chi} = 0, \qquad \tilde{T}_2 = o(1) \quad \text{as } \tilde{\chi} \to -\infty \text{ and as } \tilde{\tau} \to -\infty. \tag{44}$$

Here $\tilde{\chi}$ and $\tilde{\tau}$ are appropriately transformed space and time variables while $\tilde{T}_2$ represents the second perturbation in an expansion of the temperature.

The problem (42), (43) should be compared with that for spontaneous combustion, given by (26). Clearly they have entirely different forms.

6. Ignition by an externally generated hot spot.

Suppose some portion of an infinite combustible material is burnt very rapidly so that there is a local rise in temperature and depletion of reactant. That is, we create a hot spot (using a spark, for example) somewhat like the one that develops in auto-ignition. The hot spot can have one of two fates: either it decays by diffusion, so that after a certain time the temperature is essentially uniform once again (and constant until a homogeneous explosion occurs because of self-heating); or it acts as an ignition source, producing a deflagration wave that sweeps across and consumes the fresh material. If the hot spot is very small, temperature gradients will be very large and the resultant cooling will eliminate it. On the other hand, the results for auto-ignition suggest that a sufficiently large hot spot will ignite the material. In general, the initial-value problem that must be solved to determine the fate of a particular hot spot is difficult. We shall therefore consider a very special configuration from which plausible conclusions can be drawn quite easily.

Consider the spherically symmetric form

$$\frac{\partial T}{\partial t} - \frac{1}{r^2}\frac{\partial}{\partial r}\left(r^2\frac{\partial T}{\partial r}\right) = \mathcal{D}\left(\frac{T_b - T}{Y_f}\right)e^{-\theta/T} \qquad (45)$$

of (23). This has an exact, stationary solution

$$T = \begin{cases} T_b \\ T_f + \dfrac{Y_f r_*}{r} \end{cases} \quad \text{for } r \lessgtr r_*, \qquad (46)$$

in the limit $\theta \to \infty$, provided

$$r_* = \frac{Y_f^{3/2}\theta e^{\theta/2T_b}}{\sqrt{2\mathcal{D}T_b^2}}. \qquad (47)$$

Thus, the reaction term is asymptotically zero on either side of the flame at $r = r_*$, where the temperature gradient takes a jump

$$\left.\frac{\partial T}{\partial r}\right|_{r_*+0} = -\sqrt{\frac{2\mathcal{D}}{Y_f}}\, T_b^2 \theta^{-1} e^{-\theta/2T_b} \qquad (48)$$

(cf. the deflagration-wave solution in § 2.4).

This combustion field is now subjected to spherically symmetric disturbances. A straightforward stability analysis shows that the perturbation of the flame radius grows like e^{t/Y_f}, which corresponds to a collapsing or growing hot spot depending on whether the flame is displaced inwards or outwards initially. The result suggests that the radius (46) is critical: larger hot spots will grow and smaller ones will collapse.

LECTURE 2

Governing Equations, Asymptotics, and Deflagrations

The problem of formulating the governing equations of combustion consists, at its simplest, in characterizing the flow of a viscous, heat-conducting mixture of diffusing, reacting gases. This is a formidable task that could fill a week of lectures by itself, most of which would not be of great interest to a mathematical audience. Mindful of this, we shall limit ourselves to a description, rather than a derivation, of the simplest equations that can be brought to bear on combustion problems. Only the most important assumptions normally used to justify the equations will be discussed; for a more extensive treatment the reader is referred to Buckmaster & Ludford (1982, Chap. 1).

We shall then outline the asymptotic method on which the whole theory rests and use it to solve the basic problem of combustion: the steady propagation of a plane deflagration wave.

1. Equations for dilute mixtures. The easiest framework in which to understand the field equations is the "reactant bath". We suppose that most of the mixture consists of a single inert component (e.g. nitrogen), the properties of which determine those of the mixture (e.g. viscosity, specific heat). The reacting components (and their products) are highly diluted by immersion in this bath of inert.

Mass conservation for the mixture is always described by the single-fluid equation

$$\frac{\partial \rho}{\partial t} + \nabla \cdot (\rho \mathbf{v}) = 0, \tag{1}$$

where ρ is the density and $\mathbf{v}$ the velocity. But only for dilute mixtures is the overall momentum balance identical to that for a single homogeneous fluid, namely

$$\rho \frac{D\mathbf{v}}{Dt} = \nabla \cdot \mathbf{\Sigma}, \tag{2}$$

where

$$\mathbf{\Sigma} = -(p + \tfrac{2}{3}\kappa \nabla \cdot \mathbf{v})\mathbf{I} + \kappa[\nabla \mathbf{v} + (\nabla \mathbf{v})^T] \tag{3}$$

and p is pressure; bulk viscosity has been neglected.

A single-fluid equation for energy balance is also justified, provided account is taken of the release of heat by chemical reaction. But here additional approximations are made, based on the observation that temperatures are high and gas speeds low for a large class of combustion phenomena (excluding detonation); more precisely, a characteristic Mach number is small (typically in the range 10^{-2}–10^{-3}). Then the only significant form of energy, other than that of chemical bonding, is thermal; kinetic energy makes a negligible contribution to the energy balance. For the same reason, the conversion of kinetic energy into thermal energy by way of viscous dissipation can be ignored. Thus, when variations of the specific heat c_p with temperature are neglected, we have

$$\rho c_p \frac{DT}{Dt} - \nabla \cdot (\lambda \nabla T) - \frac{\partial p}{\partial t} = q, \tag{4}$$

where q is the heat released per unit volume of the fluid by chemical reaction; the form of q is considered later.

In addition, the assumption of small Mach number implies that spatial variations in pressure are small, so that $\partial p/\partial t$ in the energy balance (4) is due (for unconfined flames) to imposed, uniform pressure variations. We shall assume that the imposed pressure is constant, i.e. the term vanishes. The pressure term in the momentum equation (2) cannot be neglected, however; the small spatial variations are needed to account for changes in the weak velocity field. A further consequence of the virtual constancy of the pressure is that the equation of state of the mixture is Charles's law

$$\rho T = \frac{m p_c}{R} \tag{5}$$

if the inert is a perfect gas. Here m is the molecular mass of the inert, p_c the imposed constant pressure, and R the gas constant.

Consider now the individual components of the mixture, denoting the density of the i-component by ρY_i, where Y_i is the mass fraction and $i = 1, 2, \ldots, N$. The reactants and their products are convected with the gas velocity $\mathbf{v}$, diffuse relative to the inert diluent, and are consumed or generated by reaction. The diffusion laws of general mixtures are complicated, involving a diffusion matrix; but for dilute mixtures the matrix is diagonal insofar as the reactants and products are concerned, so that we may write

$$\rho \frac{DY_i}{Dt} - \nabla \cdot (\mu_{ii} \nabla Y_i) = \dot{\rho}_i \quad \text{for } i = 1, 2, \ldots, N-1. \tag{6}$$

Here $\dot{\rho}_i$ is the mass production rate per unit volume of the ith component; its precise form is considered below. The equation for the mass fraction Y_N of the inert is more complicated, but it can be obtained from the relation

$$\sum_{i=1}^{N} Y_i = 1 \tag{7}$$

instead, once the other Y_i's have been determined.

Coupling between the fluid-mechanical equations (1), (2) and the thermal-chemical equations (4), (6) occurs because of density variations. If these variations are ignored, the former may be solved for $\mathbf{v}$, which can then be substituted into the latter, a substantial simplification. Such a procedure is justified if the heat released by the reaction is small, but this is not a characteristic of combustion systems, whose main purpose is to liberate heat from its chemical bonds. For this reason, the simplified system of equations should be thought of as a model in the spirit of Oseen's approximation in hydrodynamics. However, to emphasize the mathematically rational nature of the procedure, we shall refer to the simplified system as the constant-density approximation rather than model. Phenomena whose physical basis is truly fluid-mechanical (e.g. the Darrieus–Landau instability discussed in Lecture 5) are not encompassed by this approximation, but much of importance is; it will play a central role in our discussion.

There remains the question of the contribution of the individual reactions to the heat release q and the production rates $\dot{\rho}_i$. It is possible, in principle, to consider all the reactions that are taking place between the constituents of a mixture. However, this is done but infrequently; often a complete chemical-kinetic description (i.e. how the rates depend on the various concentrations and temperature, or even whether a particular reaction takes place) is not available. Even when it is, its complexity may deter solution by anything short of massive use of computers. For these reasons, simplified kinetic schemes are normally adopted which model, in an overall fashion, the multitude of reactions.

The simplest are the one-step irreversible schemes that account for the consumption of the reactants, here taken to be just a fuel and an oxidant. If the reactants are simply lumped together as a single entity, the scheme is represented by

$$[Y_1] \to \text{products}, \tag{8}$$

where brackets denote a molecule of the component whose mass fraction is enclosed. On the other hand, if the separate identities of the fuel and oxidant are recognized, we have

$$\nu_1[Y_1] + \nu_2[Y_2] \to \text{products}; \tag{9}$$

here the ν_i are stoichiometric coefficients, specifying the molecular proportions in which the two reactants participate. We shall adopt the scheme (8) when discussing premixed combustion and (9) for diffusion flames (Lectures 8–10 only). These terms are defined at the beginning of §4.

If N_i is the number density of the ith component, so that

$$\dot{\rho}_i = m_i \dot{N}_i, \tag{10}$$

where m_i is the molecular mass of the ith component, the reaction rate ω is defined by the formula

$$\dot{N}_i = -\nu_i \omega. \tag{11}$$

It is then common to write

$$\omega = k(T) \rho^\gamma \prod_j Y_j^{\beta_j} \quad (\gamma, \beta_j \text{ positive consts.}) \tag{12}$$

for the reaction rate, an empirical formula that is suggested by a theoretical treatment of so-called elementary reactions. The product contains a single term for the scheme (8), two terms for (9). The Arrhenius law

$$k = BT^\alpha e^{-E/RT} \quad (B, \alpha, E \text{ consts.}), \tag{13}$$

which we shall adopt, is at the heart of our mathematical treatment; E is called the activation energy.

The heat release q is a consequence of the difference between the heats of formation of the products and those of the reactants, so that it is proportional to ω. Combustion is inherently exothermic, so that we shall write

$$q = Q\omega, \tag{14}$$

where $Q(>0)$ has the dimensions of energy.

2. Nondimensional equations; Shvab–Zeldovich formulation. We shall take units as follows:

$$\text{temperature } \frac{Q}{c_p \sum_j \nu_j m_j} \quad (\text{summation over 1 or 2 reactants}), \tag{15}$$

$$\text{pressure } p_c, \quad \text{density } \rho_r, \quad \text{mass flux } M_r, \quad \text{speed } \frac{M_r}{\rho_r}, \tag{16}$$

$$\text{length } \frac{\lambda}{c_p M_r}, \quad \text{time } \frac{\lambda \rho_r}{c_p M_r^2}, \quad \text{pressure variations } \frac{M_r^2}{\rho_r}. \tag{17}$$

Appropriate choices for the reference density ρ_r and the reference mass flux M_r will be made according to the problem considered. The governing equations in nondimensional form are

$$\rho T = \frac{m p_c c_p \sum_j \nu_j m_j}{\rho_r R Q}, \quad \frac{\partial \rho}{\partial t} + \nabla \cdot (\rho \mathbf{v}) = 0, \tag{18}$$

$$\rho \frac{D\mathbf{v}}{Dt} = -\nabla p + \mathscr{P}[\nabla^2 \mathbf{v} + \tfrac{1}{3}\nabla(\nabla \cdot \mathbf{v})], \tag{19}$$

$$\rho \frac{DT}{Dt} - \nabla^2 T = \Omega, \quad \rho \frac{DY_i}{Dt} - \mathscr{L}_i^{-1} \nabla^2 Y_i = \alpha_i \Omega, \tag{20}$$

where i runs from 1 to $N-1$, and

$$\mathscr{P} = \frac{\kappa c_p}{\lambda} \text{ (Prandtl number)}, \quad \mathscr{L}_i = \frac{\lambda}{\mu_{ii} c_p} \text{ (Lewis number)}, \tag{21}$$

$$\alpha_i = -\frac{\nu_i m_i}{\sum_j \nu_j m_j} \left(\text{with} \sum_i \alpha_i = -1 \right), \quad \Omega = \mathscr{D} e^{-\theta/T} \prod_j Y_j^{\beta_j}, \tag{22}$$

$$\theta = \frac{E c_p \sum_j \nu_j m_j}{QR}, \quad \mathscr{D} = D M_r^{-2}, \quad D = \frac{\lambda B Q^\alpha \rho_r^\gamma}{c_p^{1+\alpha} (\sum_j \nu_j m_j)^{\alpha-1}} \rho^\gamma T^\alpha. \tag{23}$$

Except in the case $\gamma = \alpha$, the so-called Damköhler number D is variable; it is called a number in spite of having the dimensions of M_r^2. In the context of activation-energy asymptotics ($\theta \to \infty$), only its value at a fixed temperature T_* plays a role, so that it may be considered an assigned constant.

When one of the Lewis numbers $\mathscr{L}_i$ is equal to 1, the differential operator in its equation (20b) is identical to that in (20a). We may therefore write

$$\left(\rho\frac{D}{Dt} - \nabla^2\right)\left(T - \frac{Y_i}{\alpha_i}\right) = 0, \qquad (24)$$

of which one solution is

$$T - \frac{Y_i}{\alpha_i} \equiv H_i \qquad (25)$$

a constant. If this solution is appropriate for the problem at hand, Y_i may be eliminated in favor of T, thereby reducing the number of unknowns. The linear combination (25) is known as a Shvab–Zeldovich variable; it is easier to find by virtue of satisfying the reactionless equation (24).

3. Activation-energy asymptotics. In these lectures we shall discuss a variety of combustion phenomena on the basis of equations (18)–(20). To do this we need an effective tool for dealing with the highly nonlinear reaction term Ω. Activation-energy asymptotics, used in an ad hoc fashion by the Russian school (notably Frank-Kamenetskii and Zeldovich) in the '40s, exploited in the framework of modern singular perturbation theory (but in a very narrow context) by aerothermodynamicists in the '60s, and systematically developed by Western combustion scientists in the '70s, is just such a tool.

The limit $\theta \to \infty$ is, by itself, of little interest: the definition (22b) shows that Ω vanishes. To preserve the reaction, it is necessary for $\mathscr{D}$ to become unboundedly large; i.e., we must consider a distinguished limit characterized essentially by

$$\mathscr{D} \sim e^{\theta/T_*} \qquad (26)$$

where T_* is a constant that may have to be found. The consequences of this limit then depend on the relative magnitudes of T and T_*.

For $T < T_*$, the reaction term Ω vanishes to all algebraic orders; this is known as the frozen limit. For $T > T_*$, (20) imply $\prod_j Y_j^{\beta_j} \to 0$ exponentially rapidly, so that at least one Y_i vanishes in that way, and again Ω vanishes to all orders; the equilibrium limit holds. For T no more than $O(\theta^{-1})$ from T_*, reaction takes place, usually in a thin layer called a flame sheet. Thus, with a few exceptions, the general feature of high activation energy is the absence of chemical reaction from most of the combustion field, the description of which is thereby simplified. Reaction occurs only in thin layers (spatial or temporal), whose description is also relatively simple.

4. Plane deflagration waves. We are now ready to demonstrate the efficacy of the technique that is the central theme of these lectures, by examining the fundamental problem of premixed combustion—the plane unbounded flame. But first we pause for a few words on terminology.

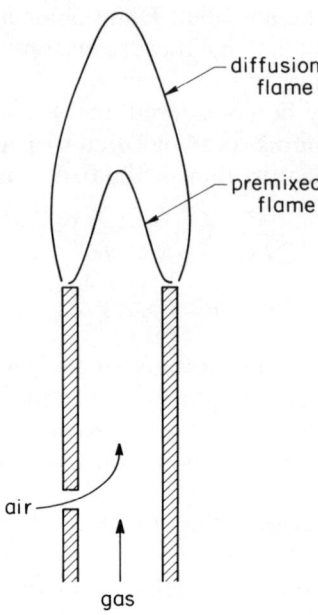

Fig. 2.1. *Bunsen burner.*

In general, two-reactant flames can be classified as diffusion or premixed. In a premixed flame the reactants are mixed and burn when the mixture is raised to a sufficiently high temperature. In a diffusion flame the reactants are of separate origin; burning occurs only at a diffusion-blurred interface.

Both kinds of flames can be produced by a Bunsen burner (Fig. 1). If the air hole is only partly open, so that a fuel-rich mixture of gas and air passes up the burner tube, a thin conical sheet of flame stands at the mouth; this is the premixed flame. Any excess gas escaping downstream mixes by diffusion with the surrounding atmosphere and burns as a diffusion flame.

Separate origins do not guarantee a diffusion flame, however. In Fig. 2 the reactants are originally separated by the splitter plate but mix before igniting. The flame spread across the oncoming flow is therefore premixed. Behind this premixed flame the remaining portions of the reactants are separate again, so that a diffusion flame trails downstream.

The plane unbounded flame of premixed combustion, the so-called (plane) deflagration wave, propagates at a well-defined speed through the fresh mixture and, accordingly, can be brought to rest by means of a counterflow. It is natural to take the mass flux of this counterflow as the representative mass flux M_r; the counterflow is not known a priori, but is to be determined during the analysis of the combustion field. Indeed, its determination is the main goal of the analysis. A choice must also be made of the reference density ρ_r; we shall take it to be that of the fresh mixture.

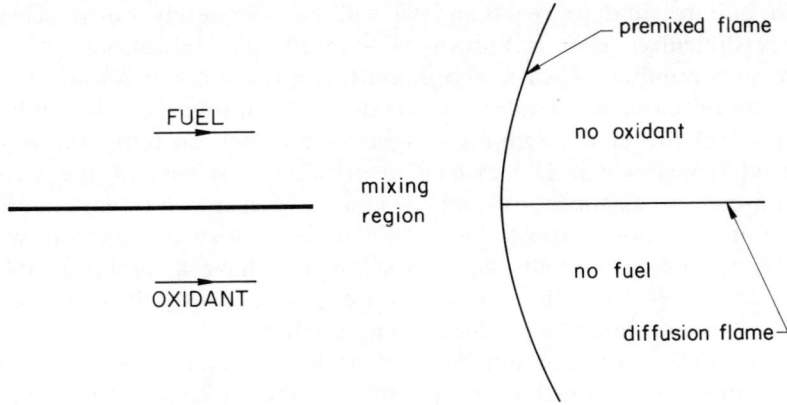

FIG. 2.2. Combustion of initially separated reactants.

The continuity equation (18b) integrates to give

$$\rho v = 1, \tag{27}$$

so that

$$\frac{dT}{dx} - \frac{d^2T}{dx^2} = \Omega. \tag{28}$$

Since there is only one reactant we shall drop the subscript 1. For $\mathscr{L} = 1$ in the corresponding equation (20b), the Shvab–Zeldovich formulation applies, showing that

$$T + Y \equiv H = T_f + Y_f \equiv T_b, \tag{29}$$

where the subscript f denotes the fresh mixture at $x = -\infty$. (Actually, H is the total enthalpy of the mixture.) Thus,

$$\Omega = \mathscr{D}(T_b - T)e^{-\theta/T} \tag{30}$$

if the most common choice $\beta_1 = 1$ is made. Equations (28) and (30) form a single equation for T, which must satisfy the boundary condition

$$T \to T_f \quad \text{as } x \to -\infty. \tag{31}$$

The requirement that all the reactant be burnt provides the final boundary condition

$$T \to T_b \quad \text{as } x \to +\infty. \tag{32}$$

Note that neither the equation of state (18a) nor the momentum equation (19) has been used; the former provides ρ once T has been found, and the latter then determines p from $v = 1/\rho$.

It is immediately apparent, since Ω does not vanish for $T = T_f$, that the problem for T cannot have a solution. The mixture at any finite location will have an infinite time to react and so will be completely burnt. This cold-boundary difficulty, as it is known, is a result of idealizations and can be resolved in a number of ways: the mixture can originate at a finite point; an appropriate initial-value problem can be defined, without the solution having a steady limit of the kind originally sought; or a switch-on temperature can be introduced below which Ω vanishes identically. It is one of the virtues of activation-energy asymptotics that it makes such resolutions unnecessary. Reaction at all temperatures below T_* (including T_f) is exponentially small, so that it takes an exponentially large time for it to have a significant effect; in other words, T_* is a switch-on temperature. Consequently, it is not necessary to discuss the cold-boundary difficulty any further.

We now seek a solution that is valid as $\theta \to \infty$. Our construction will be guided by the assumption that, in the limit, reaction is confined to a thin sheet located at $x = 0$. On either side of this flame sheet, (28) simplifies to

$$\frac{dT}{dx} - \frac{d^2T}{dx^2} = 0, \qquad (33)$$

which only has a constant as an acceptable solution behind the flame sheet ($x > 0$), exponential growth being excluded. The boundary condition (32) then shows that

$$T = T_b \quad \text{for } x > 0; \qquad (34)$$

T_b is called the adiabatic flame temperature. It follows that the temperature at the flame sheet is T_b, so that this is also the value of T_* needed to specify the distinguished limit (26).

Ahead of the flame sheet, (33) has the solution

$$T = T_f + Y_f e^x + C(\theta) e^x \quad \text{for } x < 0, \qquad (35)$$

satisfying the boundary condition (31) and making T continuous at $x = 0$ to leading order, provided C vanishes in the limit $\theta \to \infty$. No structure could be found for the flame sheet if the temperature did not satisfy this condition. A small displacement of the origin of x can absorb C which can, therefore, be set equal to zero.

Turning now to the structure, which must determine the still-unknown M_r (i.e. $\mathcal{D}$), we note that the form of Ω restricts the variations in T to being $O(\theta^{-1})$. Since the temperature gradient must be $O(1)$ to effect the transition between the profiles (34) and (35), the appropriate layer variable is

$$\xi = \theta x; \qquad (36)$$

coefficients in the layer expansion

$$T = T_b - \theta^{-1} T_b^2 \phi + \cdots \quad \text{with } \phi = \left(\frac{1}{T}\right)_1 \qquad (37)$$

are now considered to be functions of ξ. Note that the sign of ϕ has been changed from the definition (1.5), a change that will be adhered to in all subsequent lectures.

Equation (28) now shows that

$$\frac{d^2\phi}{d\xi^2} = \tilde{\mathcal{D}}\phi e^{-\phi} \quad \text{with} \quad \tilde{\mathcal{D}} = \frac{\mathcal{D}e^{-\theta/T_b}}{\theta^2}, \tag{38}$$

while matching with the solutions (34), (35) gives the boundary conditions

$$\phi = -\frac{Y_f \xi}{T_b^2} + o(1) \quad \text{as } \xi \to -\infty, \qquad \phi = o(1) \quad \text{as } \xi \to +\infty. \tag{39}$$

In order for this problem (in ϕ) to make sense, $\tilde{\mathcal{D}}$ must be $O(1)$. Then $\mathcal{D}e^{-\theta/T}$ is $O(\theta^2)$ in $x>0$, so that (28) is unbalanced unless $Y=0$ (to all orders) behind the flame sheet, consistent with the result (34). Thus, equilibrium prevails in $x>0$ even though the mixture is no hotter than the flame sheet there.

Integrating (38a) once, using the condition (39b), gives

$$\left(\frac{d\phi}{d\xi}\right)^2 = 2\tilde{\mathcal{D}}[1-(\phi+1)e^{-\phi}]. \tag{40}$$

The remaining boundary condition will then be satisfied only if

$$\tilde{\mathcal{D}} = \frac{Y_f^2}{2T_b^4}, \tag{41}$$

corresponding to the determination

$$M_r = \frac{\sqrt{2D}\, T_b^2 e^{-\theta/2T_b}}{Y_f \theta} \tag{42}$$

of the burning rate. (If D is temperature dependent it must be evaluated at the temperature T_b.)

Determination of the wave speed M_r/ρ_r is the main goal of the analysis, and rightly so. But, at the same time, the structure of the combustion field is obtained (Fig. 3). The reaction zone appears as a discontinuity in the first derivatives of T and Y, a reflection of the delta function nature of Ω in the limit $\theta \to \infty$. Ahead, the temperature rises and the reactant concentration falls as the reaction zone is approached through the so-called preheat zone. It is the

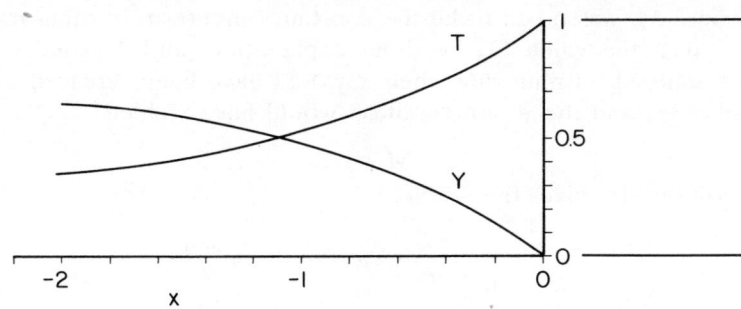

FIG. 2.3. Profiles of T and Y, drawn for $\mathcal{L}=1$ and $T_f=0.25$, $Y_f=0.75$.

preheat zone that delimits the combustion field and, therefore, defines the thickness of the flame. According to the formula (35), more than 99% of the increase in temperature from T_f to T_b is achieved in a distance 5, i.e., $5\lambda/c_p M_r$ in dimensional terms. With this definition of flame thickness, we find that hydrocarbon-air flames are about 0.5 mm thick. The thickness of the reaction zone, which is scaled by θ^{-1}, is typically 10 or 20 times smaller.

If the reaction-zone structure itself is required, (40) must be integrated to obtain ϕ as a function of ξ. The constant of integration is fixed by the boundary condition (39a).

5. Generalizations. The analysis of § 4 yields a definite value of $\mathscr{D}$; the term laminar-flame eigenvalue is often used. A similar analysis for $\mathscr{L} \neq 1$ is also possible, and then (42) is replaced by

$$M_r = \frac{\sqrt{2\mathscr{L}D}\, T_b^2 e^{-\theta/2T_b}}{Y_f \theta}. \tag{43}$$

The only change is the replacement of $\sqrt{2D}$ by $\sqrt{2\mathscr{L}D}$.

The rate at which the mixture burns is extremely sensitive to the flame temperature. If T_b changes to $T_b - \theta^{-1} T_b^2 \phi_*$, i.e., by an $O(\theta^{-1})$ amount, the burning rate changes to

$$M = M_r e^{-\phi_*/2}, \tag{44}$$

i.e., by an $O(1)$ amount. This is equivalent to a leading-order jump condition on the normal derivative $\partial T/\partial n$, namely

$$\delta\!\left(\!\left(\frac{\partial T}{\partial n}\right)^{\!2}\right) = Y_f^2 e^{-\phi_*} \tag{45}$$

when M_r is taken to have the value (42), and in this form has universal validity (i.e., is independent of the way in which ϕ_* is generated) provided the temperature gradient vanishes to leading order behind the flame sheet. The reason is that the perturbation only intrudes through the matching of ϕ at $\xi = +\infty$ (which leads to the exponential factor). This aspect of the structure problem is obscured by the analysis of the steady plane wave given in the last section, where M_r was taken to be the constant (unknown) burning rate. If M_r had been given the value (22) without explanation, and M used to denote the (dimensionless) burning rate, then x would have been replaced by Mx in the formula (35) and the jump condition would have yielded

$$M = 1. \tag{46}$$

The condition (45) gives the gradient

$$\frac{\partial T}{\partial n} = Y_f e^{-\phi_*/2} \tag{47}$$

ahead, a result that will be needed repeatedly later.

LECTURE 3

General Deflagrations

In the last lecture we examined the plane, steady, adiabatic, premixed flame and deduced an explicit formula for its speed. By a judicious choice of parameters this formula can be made to agree roughly with experiment; precision is not a reasonable goal, given the crude nature of our model. Noteworthy is the extreme sensitivity of the speed to variations in the flame temperature: an $O(1)$ change generates an exponentially large change in flame speed. Such variations in speed (caused, for example, by changes in mixture strength) are not excessive numerically (at least for fuels burnt in air), because activation energies and fractional changes in temperature are modest; but in an asymptotic analysis they present a potential obstacle to discussion of multidimensional and/or unsteady flames. Then significant variations in the flame temperature, spatial and/or temporal, can be expected and, if the sensitivity mentioned above is any guide, there will be correspondingly large variations in the flame speed. A mathematical framework in which to accommodate these is not obvious. (The first lecture dealt with special circumstances for which such variations were manageable.)

As a consequence, attempts to discuss general deflagrations have, for the most part, been limited to situations where there is an a priori guarantee that variations in the flame temperature are $O(\theta^{-1})$; then flame-speed changes are $O(1)$ and present no mathematical difficulties. Two approaches are known to provide the guarantee, and this lecture is largely devoted to their disclosure.

1. The hydrodynamic limit. At the end of § 2.4 the steady plane deflagration was found to have a thickness $5\lambda/c_p M_r$, and this may be taken as the nominal thickness of a general deflagration. We start by restricting attention to waves whose characteristic length (e.g. minimum radius of curvature) is large compared to their nominal thickness. On this length scale, such a wave is simply a surface across which jumps in temperature and density occur subject to Charles's law (as is appropriate for an essentially isobaric process).

If the ratio of the two scales is ε, then on either side of the surface the appropriate variables are

$$(\bar{x}, \bar{y}, \bar{z}, \bar{t}) = \varepsilon(x, y, z, t); \tag{1}$$

so that the governing equations (2.18b), (2.19), (2.20) become

$$\frac{\partial \rho}{\partial \bar{t}} + \bar{\nabla} \cdot (\rho \mathbf{v}) = 0, \qquad \rho \frac{D\mathbf{v}}{D\bar{t}} = -\bar{\nabla}p + \varepsilon \mathscr{P}[\bar{\nabla}^2 \mathbf{v} + \tfrac{1}{3}\bar{\nabla}(\bar{\nabla} \cdot \mathbf{v})], \tag{2}$$

$$\rho \frac{DT}{D\bar{t}} - \varepsilon \bar{\nabla}^2 T = 0, \qquad \rho \frac{DY}{D\bar{t}} - \varepsilon \mathscr{L}^{-1} \bar{\nabla}^2 Y = 0. \tag{3}$$

(We have not written the equations for components other than the single reactant $i = 1$, and the subscript 1 has been dropped.) As $\varepsilon \to 0$, we have

$$\frac{DT}{D\bar{t}} = \frac{DY}{D\bar{t}} = 0, \qquad (4)$$

i.e. constant values of T and Y are carried by the fluid particles. We conclude that

$$T = T_f, \qquad Y = Y_f \qquad (5)$$

everywhere ahead of the discontinuity surface if, as we shall suppose, these constant values are assumed by each particle at its point of origin. Likewise

$$T = T_b, \qquad Y = 0 \qquad (6)$$

everywhere behind the discontinuity since, as we shall see presently, these values are assumed by each particle as it leaves the flame. Charles's law (2.18a) now shows that ρ has the constant values ρ_f, $\rho_b = \rho_f/\sigma$ on the two sides of the discontinuity, where

$$\sigma = \frac{\rho_f}{\rho_b} = \frac{T_b}{T_f} = 1 + \frac{Y_f}{T_f} \qquad (7)$$

is the expansion ratio due to the flame. We are left the Euler's equations

$$\bar{\nabla} \cdot \mathbf{v} = 0, \qquad \rho \frac{D\mathbf{v}}{D\bar{t}} = -\bar{\nabla} p \qquad (8)$$

for an incompressible, ideal fluid, i.e. one devoid of both viscosity and heat conduction.

The two ideal-fluid regions are coupled through the jump conditions

$$\rho_f(v_{nf} + V) = \rho_b(v_{nb} + V), \qquad \mathbf{v}_{\perp f} = \mathbf{v}_{\perp b}, \qquad (9)$$

$$p_f + \rho_f(v_{nf} + V)^2 = p_b + \rho_b(v_{nb} + V)^2, \qquad T_f + Y_f = T_b; \qquad (10)$$

here V is the speed of the deflagration wave back along its normal (Fig. 1), and the subscript $\perp$ denotes the component perpendicular to $\mathbf{n}$, i.e. in the tangent plane. These conditions are derived in the same way as for a shock wave in reactionless gasdynamics, i.e. by integrating the basic equations (2.18b), (2.19), (2.20) through the flame. Indeed, the conditions (9) and (10a)

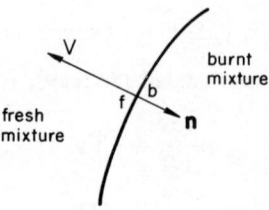

FIG. 3.1. *Notation for flame as hydrodynamic discontinuity.*

are identical to those for a shock since they follow from the same continuity and momentum equations. The requirement (10b) can also be recognized as a Rankine–Hugoniot condition, but with kinetic energy neglected and a heat-release term (Y_f) added. It follows from the combination

$$\rho \frac{D(T+Y)}{Dt} = \nabla^2(T + \mathscr{L}^{-1}Y) \tag{11}$$

of the basic equations (2.20).

As for the shock wave, these jump conditions are insufficient. If the state f immediately ahead of the wave is given, there are five scalar equations for the six unknown scalars $\rho_b(=\rho_f T_f/T_b)$, v_{nb}, $\mathbf{v}_{\perp b}$, p_b and V. In the case of a shock, another condition is imposed from outside (such as the deflection of the streamlines at a sharp body or the pressure p_b behind the wave in a shock tube). Here there is no external condition; the deficiency arises from discarding information by using only the combination (11) of the basic equations (2.20). The reaction rate then plays no role in the derivation of the jump conditions other than implying that Y vanishes for the burnt gas. Otherwise stated, the combustion inside the wave will provide information about the burning rate $\rho_f W$, i.e., the wave speed

$$W = v_{nf} + V. \tag{12}$$

Evaluation of W from a combustion analysis has often been sidestepped. Instead, hypotheses are introduced; the simplest is that W is a constant, given by the burning-rate formula (2.43) of steady, plane deflagrations. This hypothesis is justified for slowly varying flames (§ 3) when $\mathscr{L} = 1$, and we shall use it in Lecture 10. But, in general, it is not acceptable and attempts have been made (notably by Markstein) to modify it, in particular by taking into account nonplanar characteristics of the flame.

The remainder of this lecture will be preparation for the more general combustion analysis that follows in the next.

2. Governing equations for the constant-density approximation. Although the formulation can be carried through for the full equations (2.18)–(2.20), all the essential features are preserved under the assumption that density variations due to the presence of the flame are negligible. If no temperature differences are imposed on the flow, the velocity field is that of a constant-density fluid and can be calculated in advance; we shall suppose the fluid is at rest. In other words, we shall set

$$\rho = 1, \quad \mathbf{v} = 0 \tag{13}$$

in the full equations to obtain

$$\frac{\partial T}{\partial t} - \nabla^2 T = \Omega - \theta^{-1}\psi(T), \quad \frac{\partial Y}{\partial t} - \mathscr{L}^{-1}\nabla^2 Y = -\Omega \tag{14}$$

as those governing the combustion field under the constant-density approximation. (All equations (2.20b) except the first, corresponding to the single reactant, can be omitted; the subscript 1 can then be dropped.) Note that we have added a term $-\theta^{-1}\psi(T)$ to the temperature equation, representing small bulk heat loss.

If the representative mass flux M_r is chosen to be the burning rate (2.43) of the plane, steady (adiabatic) deflagration, then the reaction term becomes

$$\Omega = \mathscr{D} Y e^{-\theta/T} \quad \text{with} \quad \mathscr{D} = \frac{Y_f^2 \theta^2 e^{\theta/T_b}}{2\mathscr{L} T_b^4}. \tag{15}$$

Note that $\mathscr{L}$ is not necessarily equal to 1 in these equations: the Lewis number plays a very important role in the analysis, especially for unsteady flames. Finally, the heat-loss term is difficult to justify in a multidimensional context (radiation loss, one of the few legitimate candidates, is negligibly small unless there are solid particles such as soot in the mixture); but for quasiplane flames it can represent multidimensional effects such as losses to sidewalls.

We shall require that

$$T \to T_f, \quad Y \to Y_f \quad \text{as } x \to -\infty \tag{16}$$

and deal exclusively with situations where equilibrium prevails behind the flame sheet, i.e.

$$Y = 0 \quad \text{in the burnt gas.} \tag{17}$$

The temperature behind the flame will be close to the adiabatic flame temperature (2.29).

The constant-density approximation, on which most of the premixed flame analysis in these lectures is based, clearly provides substantial simplifications. It can be justified as a formal limit in which the heat released by the reaction becomes vanishingly small (compared to the existing thermal energy of the mixture). Small heat release can be due to either a scarcity of reactant ($Y_f \to 0$) or weak combustion ($T_f \to \infty$); by confining ourselves to dilute mixtures, we have already assumed the former. The relevant parameter is the expansion ratio (7); asymptotic expansions in $\sigma - 1$ provide a formal basis for the approximation.

3. Slow variations with loss of heat. As an introduction to the more complicated analysis of multidimensional flames consider first a plane flame sheet (Fig. 1), looking like the adiabatic deflagration studied in § 2.4 but moving unsteadily because of local fluctuations in T and Y (represented by the time derivatives). In general, the flame speed can be defined in terms of the mass flux of the mixture through the sheet. (This is a well-defined concept in the limit $\theta \to \infty$; for θ finite there is no natural definition, except when the combustion field is steady in some moving frame.) For the constant-density approximation adopted here, the speed is just

$$V = -\dot{x}_*(t). \tag{18}$$

Note that, since the speed is not defined for finite θ, to expand it in the subsequent asymptotic analysis would be a futile gesture.

Suppose now that changes in the flame speed occur on an $O(\theta)$ time scale, i.e. that

$$\tau = \frac{t}{\theta} \qquad (19)$$

is the appropriate (slow) time variable to describe them. Then, for an observer moving with the flame sheet, the combustion field is quasisteady to leading order (i.e. steady for $t = O(1)$). The temporal variations, along with heat loss, create $O(\theta^{-1})$ perturbations, and hence generate only $O(\theta^{-1})$ variations in the flame temperature.

When the coordinate

$$n = x - x_*(t), \qquad (20)$$

based on the flame sheet, is introduced, the basic equations (14) become

$$\theta^{-1}\frac{\partial T}{\partial \tau} + V\frac{\partial T}{\partial n} - \frac{\partial^2 T}{\partial n^2} + \theta^{-1}\psi = -\theta^{-1}\frac{\partial Y}{\partial \tau} - V\frac{\partial Y}{\partial n} + \mathscr{L}^{-1}\frac{\partial^2 Y}{\partial n^2} = \Omega. \qquad (21)$$

These govern the motion of what is known as a slowly varying flame (SVF). To integrate the equations it is first necessary to say something about the flame temperature. To leading order, we have

$$V\frac{\partial (T+Y)}{\partial n} = \frac{\partial^2 (T+\mathscr{L}^{-1}Y)}{\partial n^2} \qquad (22)$$

everywhere; so that, on integrating from $n = -\infty$ to $0+$ and using the boundary conditions (16), (17), we have

$$V(T_* - T_f - Y_f) = \left.\frac{\partial T}{\partial n}\right|_{0+}, \qquad (23)$$

where T_* is the leading-order temperature at the flame. Since the derivative vanishes (as will be seen immediately), we conclude that

$$T_* = T_b, \qquad (24)$$

the adiabatic flame temperature (2.29).

In view of the requirements (16), (17) the solution of (21) in the frozen region ahead of the flame sheet is

$$T = T_f + Y_f e^{Vn}, \qquad Y = Y_f(1 - e^{\mathscr{L}Vn}) \quad \text{for } n<0, \qquad (25)$$

correct to leading order. To the same order behind the flame sheet, T is constant (hence showing that the derivative in the result (23) is zero, as anticipated); to one more term we find

$$T = T_b - \theta^{-1}[V^{-1}\psi(T_b)n + T_b^2\phi_*(\tau)], \qquad Y = 0 \quad \text{for } n>0 \qquad (26)$$

by writing $T = T_b$ in the θ^{-1} terms of the temperature equation. Here ϕ_*, representing the perturbed flame temperature, is as yet unknown.

The structure problem for the reaction zone determines ϕ_* as a function of V. This problem has already been discussed in §2.5, where the expression (2.47) for the temperature gradient just ahead of the flame sheet was developed. The same gradient can be calculated from the result (25a), leading to the relation

$$V = e^{-\phi_*/2}, \tag{27}$$

which is equivalent to (2.44).

Clearly, there is the same temperature sensitivity as for steady adiabatic deflagrations, as expected. Moreover, for the latter the perturbation ϕ_* vanishes and $V = 1$, which confirms the burning-rate formula (2.43).

Another relation between ϕ_* and V comes from calculating the change in enthalpy of the mixture between its fresh and burnt states. For that purpose, we rewrite (22) correct to $O(\theta^{-1})$ before integrating it as before, to obtain

$$\theta^{-1} \int_{-\infty}^{0+} \frac{\partial}{\partial \tau}(T+Y)\,dn + [V(T+Y)]_{-\infty}^{0+} + \theta^{-1} \int_{-\infty}^{0+} \psi\,dn = \left[\frac{\partial T}{\partial n} + \mathscr{L}^{-1}\frac{\partial Y}{\partial n}\right]_{-\infty}^{0+}. \tag{28}$$

The integrals can be evaluated to leading order by means of the formulas (25); we find

$$\int_{-\infty}^{0+} \frac{\partial}{\partial \tau}[Y_f(e^{Vn} - e^{\mathscr{L}Vn})]\,dn = Y_f \dot{V} \int_{-\infty}^{0+} n(e^{Vn} - e^{\mathscr{L}Vn})\,dn = -Y_f(1-\mathscr{L}^{-1})V^{-2}\dot{V},$$

$$\int_{-\infty}^{0+} \psi(T_f + Y_f e^{Vn})\,dn = \left[\int_0^\infty \psi(T_f + Y_f e^{-v})\,dv\right] V^{-1},$$

where the dot is used to signify rate of change on the τ-scale. The formulas (26) and boundary conditions (16) enable the remaining terms to be calculated; we have

$$[V(T+Y)]_{-\infty}^{0+} = V(T_* - T_b) = -\theta^{-1}T_b^2 V\phi_*,$$

$$\left[\frac{\partial T}{\partial n}\right]_{-\infty}^{0+} = -\theta^{-1}\psi(T_b)V^{-1}, \quad \left[\frac{\partial Y}{\partial n}\right]_{-\infty}^{0+} = 0.$$

The equation (28), in which all terms have now been evaluated to $O(\theta^{-1})$, therefore gives

$$\phi_* = \Psi V^{-2} - bV^{-3}\dot{V} \quad \text{with } b = \frac{Y_f(1-\mathscr{L}^{-1})}{T_b^2}; \tag{29}$$

here

$$\Psi = \frac{\psi(T_b) + \int_0^\infty \psi(T_f + Y_f e^{-v})\,dv}{T_b^2}, \tag{30}$$

the two terms representing heat lost to the burnt mixture and through the sidewalls ahead of the flame sheet, respectively.

By eliminating ϕ_* between the two relations (27) and (29), we obtain an equation for V, namely

$$b\dot{V} = V^3 \ln V^2 + \Psi V. \tag{31}$$

The only difference when the constant-density approximation is not used is a more complicated formula for b. The crucial property

$$b \lessgtr 0 \quad \text{accordingly as} \quad \mathscr{L} \lessgtr 1 \tag{32}$$

is unaffected, however. Note that the SVF is not a solution of the general initial-value problem (only the value of V may be prescribed at $\tau = 0$); it merely describes the subsequent behavior of any flame that survives development on the t-scale. Thus, a prediction of instability is reliable, but one of stability is not, since the flame may have already lost stability during its evolution on the t-scale.

Consider first the steady state (Fig. 2) determined by setting $\dot{V} = 0$ in the evolution equation (31), i.e.

$$V = 0 \quad \text{or} \quad V^2 \ln V^2 + \Psi = 0. \tag{33}$$

On the first of these $\phi_* = +\infty$, so that the perturbation analysis breaks down; the corresponding nonuniformity has never been treated. The second curve provides two solution branches so long as the heat loss is not too large, i.e. Ψ is less than e^{-1} ($=0.368$); the adiabatic flame speed $V = 1$ is attained on the upper branch as $\Psi \to 0$, so that it is plausible to suppose that this is the physically relevant one. No solution exists for $\Psi > e^{-1}$: steady combustion cannot be sustained if the heat loss is too large, any existing flame being

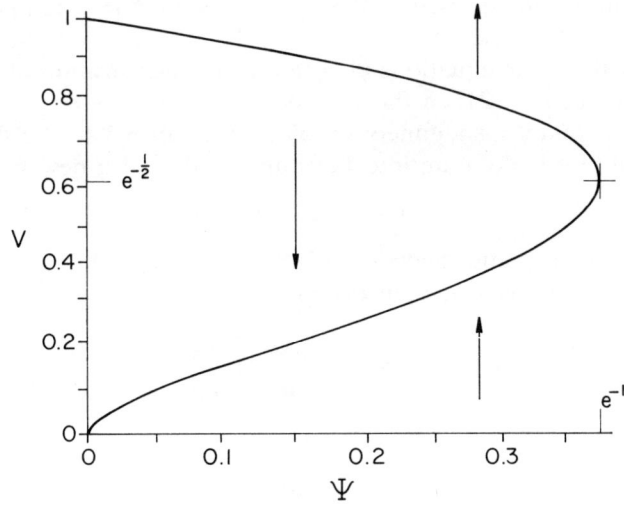

FIG. 3.2. *Steady flame speed V versus heat-loss parameter Ψ. Arrows show direction in which speed changes for $\mathscr{L} = 1$.*

quenched. It is interesting that the speed of the flame at quenching, namely $e^{-1/2}$ (=0.607) times its adiabatic value, is completely independent of the nature of the heat loss, i.e. the form of the function ψ. The quenching phenomenon provides a qualitative explanation of the Davy safety lamp: the wire gauze surrounding the flame is an effective heat sink, preventing the propagation of the flame beyond its confines.

Equation (31) describes the evolution of plane SVFs. When $\mathscr{L} = 1$, b is zero and there is no evolution: equidiffusion prevents any variation on the τ-scale. In fact, since the equation is asymptotic, there is no evolution when $\mathscr{L}$ is close to 1, i.e.

$$\mathscr{L}^{-1} = 1 - \frac{l}{\theta} \quad \text{with } l = O(1). \tag{34}$$

But then a treatment on the t-scale is possible in certain circumstances, leading to the near-equidiffusion flame (NEF) discussed later.

An immediate consequence of the evolution for $\mathscr{L} > 1$ ($b > 0$) is that the flame is unstable: any deviation of V from its value on the upper branch of the curve in Fig. 2 is amplified. The same conclusion cannot be drawn for $\mathscr{L} < 1$, but this is a consequence of considering planar disturbances only. Lecture 5 will examine the linear stability of plane deflagration waves in complete detail, and find that plane SVFs are unstable to nonplanar disturbances for $\mathscr{L} > 1$. Thus, the SVFs are unstable for all values of $\mathscr{L}$, which decreases their value as a class of solutions (but does not eliminate them).

4. Multidimensional flames. Consider now situations in which the flame sheet, in addition to being unsteady, moves in a nonplanar fashion. The goal is to find conditions under which the variations in flame temperature, both temporal and spatial, are $O(\theta^{-1})$ at most. To that end we shall perform an integration of the basic equations (14) that is a generalization of the one done on their plane version (21) in the last section.

The x-axis is taken instantaneously along the normal to the flame sheet at the point of interest (pointing into the burnt gas), and a new variable

$$n = x - F(0, 0, t) \tag{35}$$

is introduced, as for plane sheets (cf. (20)); here $F(y, z, t)$ denotes the position of the sheet. Equations (14) then become

$$\frac{\partial T}{\partial t} + V \frac{\partial T}{\partial n} - \frac{\partial^2 T}{\partial n^2} - \nabla_\perp^2 T + \theta^{-1} \psi(T) = -\frac{\partial Y}{\partial t} - V \frac{\partial Y}{\partial n} + \mathscr{L}^{-1} \frac{\partial^2 Y}{\partial n^2} + \mathscr{L}^{-1} \nabla_\perp^2 Y = \Omega \tag{36}$$

where

$$V = -\dot{F}(0, 0, t) \tag{37}$$

is the speed of the sheet back along its normal at the instant considered and, as in § 1, the subscript $\perp$ denotes the component perpendicular to **n**.

Equation (36a) is now integrated with respect to n from $-\infty$ to $0+$, thereby yielding

$$\int_{-\infty}^{0+} \frac{\partial}{\partial t}(T+Y)\,dn + [V(T+Y)]_{-\infty}^{0+} + \theta^{-1}\int_{-\infty}^{0+} \psi\,dn \qquad (38)$$

$$= \left[\frac{\partial T}{\partial n} + \mathscr{L}^{-1}\frac{\partial Y}{\partial n}\right]_{-\infty}^{0+} + \int_{-\infty}^{0+} \nabla_\perp^2(T+\mathscr{L}^{-1}Y)\,dn,$$

which should be compared to (28). Certain terms can be evaluated almost as there; thus,

$$[V(T+Y)]_{-\infty}^{0+} = V(T_* - T_b), \quad \left[\frac{\partial T}{\partial n}\right]_{-\infty}^{0+} = \frac{\partial T}{\partial n}\bigg|_{0+}, \quad \left[\frac{\partial Y}{\partial n}\right]_{-\infty}^{0+} = 0,$$

so that we may write

$$V(T_* - T_b) = \frac{\partial T}{\partial n}\bigg|_{0+} + \int_{-\infty}^{0+}\left[\nabla_\perp^2(T+\mathscr{L}^{-1}Y) - \frac{\partial H}{\partial t}\right]dn - \theta^{-1}\int_{-\infty}^{0+}\psi\,dn. \qquad (39)$$

This expresses the deviation of the flame temperature T_* from its adiabatic value T_b in terms of the heat lost to the burnt mixture, the transverse diffusion of heat and reactant up to the flame sheet, the temporal variations in enthalpy H of the mixture ahead of the flame sheet, and the heat loss up to the flame sheet.

If deviations of T_* from T_b are to be $O(\theta^{-1})$, the right side of (39) must be of the same order. This is guaranteed when the terms in $\partial/\partial n$, $\nabla_\perp^2$, and $\partial/\partial t$ are made separately small, a step that can be taken in two different ways. One way is to confine attention to disturbances of a steady, plane deflagration that vary over times and distances $O(\theta)$. These SVFs are a generalization of the ones introduced in the last section, where only temporal variations were considered. The second way is suggested by the ineffectiveness of the SVF analysis for $\mathscr{L}$ close to 1. In the distinguished limit (34), equation (36a) becomes

$$\frac{\partial H}{\partial t} + V\frac{\partial H}{\partial n} - \left(\frac{\partial^2}{\partial n^2} + \nabla_\perp^2\right)H = \theta^{-1}\left[l\left(\frac{\partial^2}{\partial n^2} + \nabla_\perp^2\right)Y - \psi(T)\right], \qquad (40)$$

of which

$$H = H_f + O(\theta^{-1}) \quad \text{(everywhere)} \qquad (41)$$

is one solution. For the corresponding class of solutions, called near-equidiffusion flames (NEFs),

$$\frac{\partial T}{\partial n}\bigg|_{0+} = \frac{\partial H}{\partial n}\bigg|_{0+}, \quad \nabla_\perp^2(T+\mathscr{L}^{-1}Y) = \nabla_\perp^2 H + O(\theta^{-1}), \quad \frac{\partial H}{\partial t} \qquad (42)$$

are all $O(\theta^{-1})$, so that the right side of equation (39) is of that order.

It should be emphasized that SVFs and NEFs are restricted classes of solutions, identified by sufficient (but not necessary) conditions for the flame-temperature variations to be $O(\theta^{-1})$, itself a sufficient condition for the efficacy of our asymptotic method. While these classes may be the only general ones, special circumstances make it possible to treat other premixed flames. Lack of time prevents our discussing the most important of these, namely the spherical (premixed) flame: symmetry ensures that the temperature does not vary at all over its flame sheet, so that it need not be either an SVF or an NEF. (Nevertheless, for certain parameter values it is an SVF and for others an NEF.)

In the next lecture, the equations governing the SVF and the NEF will be derived and then solved for a basic nonuniform velocity field: stagnation-point flow.

LECTURE 4

SVFs and NEFs

For want of a complete analysis of multidimensional flows in pre-asymptotic days, it was natural to try to identify special characteristics that play particularly important roles in the understanding of flame behavior. Flame speed and temperature are examples of such characteristics that have already been identified; a more subtle characteristic, introduced by Karlovitz, is flame stretch. We shall start by discussing this concept, so as to have it available when the later analysis is reached.

1. Flame stretch. To define this (or for that matter flame speed) in an unambiguous fashion, we must first define a flame surface, i.e. a surface characterizing the location of the reaction. For large activation energy the reaction zone is such a surface when viewed on the scale of the preheat zone, since it then collapses into the flame sheet. If the flame can be viewed as a hydrodynamic discontinuity, as in § 3.1, then the discontinuity itself is a flame surface. In either case, a flow velocity is defined on each side of the surface, such that $\mathbf{v}_\perp$ is continuous across the surface.

Consider a point that remains on the (moving) flame surface but travels along it with the velocity $\mathbf{v}_\perp$. The set of such points forming a surface element of area S will, in general, be deformed by the motion, so that S will vary with time (Fig. 1). If S increases, the flame is said to be stretched; if S decreases, the flame experiences negative stretch and is said to be compressed. A measure of the stretch is the proportionate rate of change

$$K \equiv S^{-1} \frac{dS}{dt}, \qquad (1)$$

known as the Karlovitz stretch. Note that d/dt is not a material derivative; the fluid particles in the surface element change. The points advance with the flame surface, i.e. at the speed V and not v_n.

The deformation of the surface element consists of two parts corresponding to the motions with speed V back along the normal and with velocity $\mathbf{v}_\perp$ tangentially. The first, known as dilatational stretch, is found to be κV, where κ is the first (or mean) curvature of the surface, taken to be positive when the surface is concave towards the burnt gas; the other, known as extensional stretch, is $\boldsymbol{\nabla}_\perp \cdot \mathbf{v}_p$, where $\mathbf{v}_p$ is the tangential component of velocity at neighboring points projected onto the tangent plane at the point of interest. Since

$$\boldsymbol{\nabla}_\perp \cdot \mathbf{v}_p = \boldsymbol{\nabla}_\perp \cdot \mathbf{v}_\perp + \kappa v_n, \qquad (2)$$

the Karlovitz stretch is

$$K = \boldsymbol{\nabla}_\perp \cdot \mathbf{v}_\perp + \kappa(V + v_n) = \boldsymbol{\nabla}_\perp \cdot \mathbf{v}_{\perp f} + \kappa W \quad \text{with} \quad W = V + v_{nf}. \qquad (3)$$

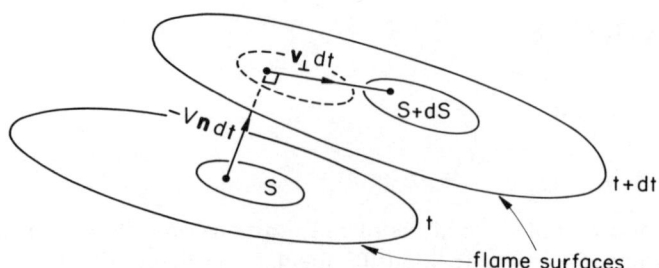

FIG. 4.1. *Flame stretch.*

Thickness is another concept characterizing a flame that can be treated as a hydrodynamic discontinuity. In § 3.1 the nominal thickness $5\lambda/c_p M_r$ was introduced, but here we need a local, instantaneous definition. It is natural to replace M_r with M, and this is found to be appropriate: decay of the temperature in the preheat zone takes place over distances proportional to M^{-1} (cf. (3.25)). In turn $M = \rho(V + v_n) = \rho_f W$. In the context of the constant density approximation, the subscript f may be discarded in these formulas and ρ_f may be replaced by 1.

Introduction of the notion of thickness leads to the concept of flame volume associated with a surface element; this is proportional to

$$\Delta = \frac{S}{W}. \tag{4}$$

Just as changes in surface element led to the concept of Karlovitz stretch, so changes in volume lead to the voluminal stretch

$$B \equiv \Delta^{-1} \frac{d\Delta}{dt} = K - W^{-1} \frac{dW}{dt} \tag{5}$$

introduced by Buckmaster. The stretch B arises naturally in our consideration of SVFs, to which we turn next.

2. The basic equation for SVFs. The discussion now focuses on the combustion field, i.e. the internal structure of the discontinuity. To confine attention to changes that occur over times and distances $O(\theta)$, we make the transformation

$$(y, z, t) = \theta(\eta, \zeta, \tau). \tag{6}$$

Correct to $O(\theta^{-1})$, the governing equations (3.14) reduce to the form (3.21) with $\partial^2/\partial n^2$ replaced by $\partial^2/\partial n^2 - \theta^{-1} \kappa \partial/\partial n$, where n is the local normal at the instant considered (cf. below). As a consequence, the results (3.25), (3.26) are still valid provided ϕ_* is allowed to depend on η, ζ as well as on τ. Of course, V does likewise (in spite of the apparent contradiction (3.37)).

One relation between V and ϕ_* is given by the universal result (3.27). The second comes from the generalization (3.39) of the enthalpy balance used in

§ 3.3. Comparison of the balances (3.28) and (3.39) shows that we have to deal with just one new term, namely

$$\int_{-\infty}^{0+} \nabla_\perp^2 (T + \mathscr{L}^{-1} Y)\, dn, \tag{7}$$

and the continued validity of the formulas (3.25), (3.26) ensures that the evaluations of corresponding terms in the two balances are the same. At first glance the term (7) appears to be $O(\theta^{-2})$, and hence negligible, because the operator $\nabla_\perp$ is $O(\theta^{-1})$. It is important to realize, however, that T, Y are given by the formulas (3.25), (3.26) only when n has its local meaning. The curvature of the flame sheet, from which n is measured, thereby provides a contribution $-\theta^{-1}\kappa\, \partial/\partial n$ to $\nabla_\perp^2$, so that the term (7) becomes

$$-\theta^{-1}\kappa(T_b - T_f - \mathscr{L}^{-1} Y_f) = \theta^{-1}\kappa(\mathscr{L}^{-1} - 1) Y_f; \tag{8}$$

here κ is the θ-multiplied first curvature of the flame sheet. The second relation between ϕ_* and V is therefore

$$\phi_* = \Psi V^{-2} - bV^{-3}\dot{V} + bV^{-1}\kappa, \tag{9}$$

which should be compared with the plane result (3.29).

Elimination of ϕ_* between the two relations now gives the basic equation

$$b(\dot{V} - V^2\kappa) = V^3 \ln V^2 + \Psi V \tag{10}$$

of an SVF, which should be compared with the plane result (3.31). It can be recast in terms of the stretch concepts introduced in §1 by noting that

$$W = V, \qquad K = \kappa V \tag{11}$$

for the quiescent flow (3.13) on which our analysis has been based. Thus,

$$b\!\left(W^{-1}\frac{dW}{d\tau} - K\right) \equiv -bB = W^2 \ln W^2 + \Psi, \tag{12}$$

where K and B, the proportionate rates of change in surface and volume elements, are measured on the slow time scale τ (just as κ is measured on the $O(\theta)$ distance scale), and b has the definition (3.29a). In this form, the equation is valid for an arbitrary flow field, not just the quiescent one that we have considered for the sake of simplicity and for which the superficial stretch is purely dilatational. When the constant-density approximation is abandoned, there are two modifications or, rather, reinterpretations. The parameter b becomes a more complicated function of $\mathscr{L}$, but still has the property (3.32). In addition, the equation is then only valid for the hydrodynamic discontinuity, not for the flame sheet; here it is valid for either, because the velocity field does not change through the flame.

The basic equation (12) for SVFs does not, in general, determine the wave speed W directly; it is a (complicated) partial differential equation for the shape of the moving surface. Only for $\mathscr{L} = 1$ (i.e. $b = 0$) does it reduce to an equation for W; in particular, $W = 1$ in the absence of heat loss. For plane deflagrations, there is no superficial stretch ($K = 0$) but there is voluminal stretch ($B = -V^{-1}\dot{V}$), due to changes in flame thickness, and (3.31) can be interpreted in terms of it.

3. The effect of stretch on SVFs. We have introduced the concept of stretch because of the importance attached to it in the past thirty years. Far-reaching use has been made of it as an intuitive tool in the prediction and explanation of flame behavior, particularly of quenching. In essence, the claim is that stretching a flame causes it to decelerate, and stretching it too much will extinguish it. While the claim has matured with time, its essence persists. However, until SVFs were identified and their connection with stretch was discovered, the claim was no more than conjecture: now we can deduce the effect of stretch on flame speed from the basic equation (12), at least for SVFs. More about stretch, in the context of NEFs, will appear in § 10.5.

To be sure, the stretch involved in (12) is B and not K; but, if the thickness does not vary, there is no distinction. (An example is the stagnation-point flow treated next.) Consider first adiabatic conditions, i.e. $\Psi = 0$. From Fig. 2, which shows a plot of $W^2 \ln W^2$ versus W^2, it is clear that, for $b > 0$ (i.e. $\mathscr{L} > 1$), positive values of B correspond to values of W less than 1, and that there is no value of W for

$$B > e^{-1} b^{-1} > 0. \tag{13}$$

The effect of stretch is indeed as conjectured, provided the Lewis number is bigger than 1. But, for a Lewis number less than 1, this is not so: for $b < 0 (\mathscr{L} < 1)$, positive values of B correspond to values of W greater than 1, and there is no limit; the flame accelerates when stretched, and can tolerate

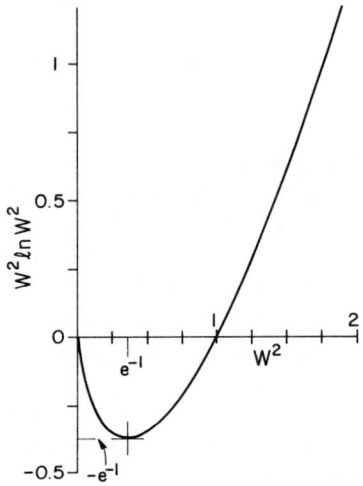

FIG. 4.2. *Graph determining effect of voluminal stretch on flame speed.*

any stretch. In fact, deceleration is associated with negative stretch (compression), of which the flame cannot tolerate too much: for

$$B < e^{-1}b^{-1} < 0, \tag{14}$$

there is no solution.

When there is heat loss, i.e. for $\Psi \neq 0$, the conditions (13), (14) are replaced by

$$B \gtrless (e^{-1} - \Psi)b^{-1} \quad \text{accordingly as } \mathscr{L} \gtrless 1; \tag{15}$$

the heat loss helps to extinguish the stretched or compressed flame. In fact, when the loss is large enough (namely for $\Psi > e^{-1}$) no stretch or compression is required at all, a result in accord with that for steady plane deflagrations in § 3.3. Moreover, the extinction speed $e^{-1/2}$ obtained there is now seen to have general validity for SVFs.

These conclusions about extinction are only of interest if it is known that stretch of the required amount (positive or negative) can be applied to a flame. It is conceivable that, when there is insufficient heat loss for extinction, the flame can always adapt to the flow conditions so as to avoid being extinguished. The stagnation-point flow considered next shows that this is not so.

We add

$$X = \frac{x}{\theta} \tag{16}$$

to the slowly varying coordinates (6), and now take axes as shown in Fig. 3.

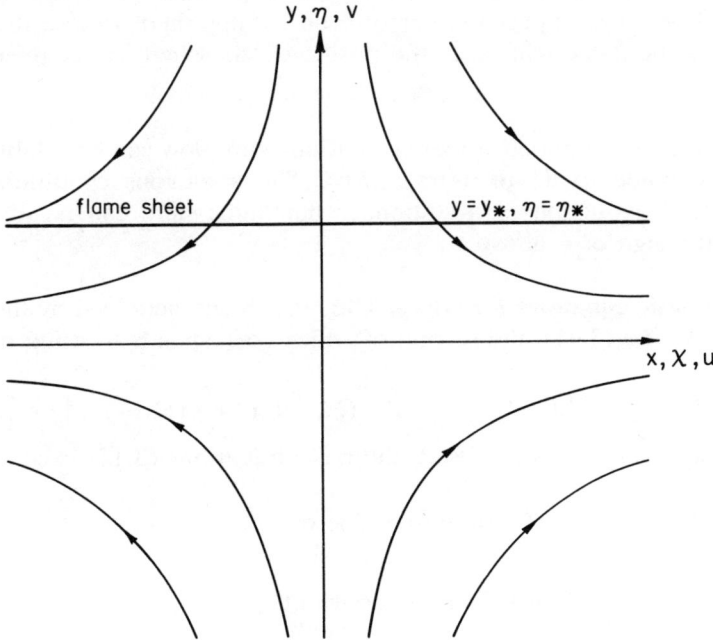

FIG. 4.3. Notation for SVF and NEF in a stagnation-point flow.

The stagnation-point flow is then

$$\mathbf{v} = \varepsilon(\chi - \eta), \tag{17}$$

where $\varepsilon > 0$ is the rate of strain. (No confusion will result from having used ε in a different way in § 3.1.) While there is a whole family of solutions of the basic equation (12), we shall concentrate on the possibility of a stationary flat flame, located at

$$\eta = \eta_* \quad \text{(say)}. \tag{18}$$

Assuming such a flame exists, we have

$$W = \varepsilon \eta_*, \quad \kappa = 0, \quad v_\perp = \varepsilon \chi, \tag{19}$$

so that its thickness is constant and

$$B = K = \frac{\partial(\varepsilon \chi)}{\partial \chi} = \varepsilon \tag{20}$$

according to the definition (3). The stretch, whether voluminal or superficial, is just the strain rate. We conclude from the result (15) that, under adiabatic conditions, a stationary flat flame exists for all (positive) ε when $\mathscr{L}$ is less than one, but not for

$$\varepsilon > e^{-1}b^{-1} \tag{21}$$

when $\mathscr{L}$ is greater than one.

There is no practical difficulty to increasing ε; just the speed of the incident stream has to be raised. As the strain rate is gradually increased up to the values (21), we may expect the flame to be extinguished. Before these values are reached the flame will lie at the position (18), where η_* is given by

$$2\varepsilon \eta_*^2 \ln (\varepsilon \eta_*) + b = 0. \tag{22}$$

Under certain conditions, a rear stagnation-point flow can be established and a flame be made to lie in it (cf. § 7.6). The analogous conditions for the existence and extinction of a stationary flat flame here can be obtained by changing the sign of ε above.

4. The basic equations for NEFs. The NEF is characterized by the requirements (3.34) and (3.41), the second of which corresponds to using the expansions

$$T = T_0 + \theta^{-1} T_1 + \cdots, \qquad Y = (H_f - T_0) + \theta^{-1}(H_1 - T_1) + \cdots. \tag{23}$$

When these assumptions are used, the basic equations (3.14) become

$$\frac{\partial T}{\partial t} = \nabla^2 T \quad \text{on either side of the flame sheet,} \tag{24}$$

$$\frac{\partial h}{\partial t} = \nabla^2 h + l \nabla^2 T \quad \text{everywhere,} \tag{25}$$

if bulk heat loss is negligible; here T stands for T_0, and H_1 has been replaced by h. Boundary and initial conditions must be consistent with the assumption that H is constant to leading order, emphasizing once more that NEFs are a restricted class of solutions.

Ahead of the flame sheet the full equations (24), (25) hold; but in the burnt gas the assumption of equilibrium leads to

$$T = T_b, \qquad \frac{\partial h}{\partial t} = \nabla^2 h \qquad (26)$$

there, the temperature perturbation accounting for the whole of h. The solutions on the two sides must be linked by jump conditions, to be derived next.

These conditions are deduced by analysis of the reaction-zone structure, a question that was addressed in §§ 2.4 and 2.5. First, the very existence of a structure requires

$$\delta(T) = 0 \quad \text{with } \delta(\cdot) = (\cdot)_{0+} - (\cdot)_{0-}; \qquad (27)$$

then, when $\partial T/\partial n = 0$ for $n = 0+$ (as here), the structure gives

$$\left.\frac{\partial T}{\partial n}\right|_{0-} = Y_f e^{-\phi_*/2}, \qquad (28)$$

where $-T_b^2 \phi_*$ is the flame-temperature perturbation, i.e. the value of h at the flame sheet. The remaining jump conditions

$$\delta(h) = 0, \qquad \delta\left(\frac{\partial h}{\partial n}\right) = l \left.\frac{\partial T}{\partial n}\right|_{0-} \qquad (29)$$

come from integrating (25) through the reaction zone and matching the result with the combustion fields outside.

The equations governing NEFs have been developed under the assumption (13b), i.e. a quiescent mixture. When the mixture is in motion they must be replaced by

$$\frac{DT}{Dt} = \nabla^2 T, \qquad \frac{Dh}{Dt} = \nabla^2 h + l\nabla^2 T \qquad (30)$$

ahead of the flame sheet, and

$$T = T_b, \qquad \frac{Dh}{Dt} = \nabla^2 h \qquad (31)$$

behind. The system (27)–(31) defines a free-boundary problem (elliptic if steady) of the fourth order, with the flame sheet as the moving boundary. Solution is a formidable question, tackled in three ways that may be listed as follows:
  (i) small perturbations,
  (ii) numerical integration,
  (iii) special geometries.

Stability considerations fall under (i); NEFs are prominent in the stability Lectures 5, 6, and 7. (Unlike SVFs they are stable for certain values of the Lewis number.) The numerical work under (ii) has dealt only with a parabolic limit of the elliptic problem; some resulting Stefan problems are considered in Lecture 10. An example of (iii), stagnation-point flow, is discussed in the next section, where the effect of stretch will be examined once more.

The discussion of general deflagrations started in Lecture 3 with a consideration of hydrodynamic discontinuities, i.e. waves whose length scale is large compared to their thickness (as represented by the parameter ε in § 3.1). The need to know the wave speed then led to an examination of the flame structure, and the uncovering of SVFs and NEFs as classes of solutions that could be handled by the asymptotics. The SVF is an acceptable structure for the interior of the hydrodynamic discontinuity if

$$\theta = O(\varepsilon^{-1}), \tag{32}$$

since then the undulations of the flame sheet follows those of the discontinuity.

No demand of the type (32) is made of NEFs; the activation energy is independently large. In other words, NEFs exist whether hydrodynamic discontinuities do or not. If a NEF can be viewed as a hydrodynamic discontinuity, it corresponds to a solution with variations on the scale ε^{-1} (other than in the n-direction). To leading order it must, therefore, be a steady, plane deflagration traveling at the adiabatic speed: $W = 1$ in the jump conditions (3.9), (3.10), (3.12).

NEFs are most useful when they cannot be viewed as hydrodynamic discontinuities; witness what we shall have to say about them from now on. Their power is evident in the stability considerations of Lecture 5.

5. NEFs near a stagnation point. The problem is sketched in Fig. 3. The velocity field is

$$\mathbf{v} = \varepsilon(x, -y), \tag{33}$$

where ε is the rate of strain, so that equations (30), (31) have solutions for which T and h are functions of y only. The combustion field can be stratified with the flame flat, as for an SVF. If the flame sheet is located at

$$y = y_*, \tag{34}$$

then

$$W = \varepsilon y_*, \quad \kappa = 0, \quad v_\perp = \varepsilon \chi, \tag{35}$$

and the Karlovitz stretch

$$K = \frac{\partial(\varepsilon x)}{\partial x} = \varepsilon \tag{36}$$

is just the strain rate.

If the wall $y = 0$ is a thermal insulator, or if there is an identical opposing jet in $y < 0$, the boundary condition

$$\frac{\partial h}{\partial y} = 0 \quad \text{at } y = 0 \tag{37}$$

must be applied. (The leading-order temperature (31a) satisfies the corresponding condition automatically.) The requirement (37) is also satisfied when the flow is uniform, the flame being then plane with reaction zone at $y = y_*$. The only role of the wall is to change the uniform flow into one that stretches the flame; heat-loss effects in addition to this geometrical role are prevented by the condition (37).

Behind the flame sheet

$$\frac{d^2 h}{dy^2} + \varepsilon y \frac{dh}{dy} = 0, \tag{38}$$

so that

$$h = -T_b^2 \phi_* \quad \text{for } 0 < y < y_* \tag{39}$$

in view of the condition (37). Ahead of the flame sheet

$$\frac{d^2 T}{dy^2} + \varepsilon y \frac{dT}{dy} = 0, \quad \frac{d^2 h}{dy^2} + \varepsilon y \frac{dh}{dy} = l\varepsilon y \frac{dT}{dy} \quad \text{for } y_* < y < \infty \tag{40}$$

while

$$T \to T_f, \quad h \to 0 \quad \text{as } y \to \infty. \tag{41}$$

At the flame sheet itself, the jump conditions (27)–(29) require

$$T = T_b, \quad h = -T_b^2 \phi_*, \quad \frac{dT}{dy} = -l^{-1} \frac{dh}{dy} = -Y_f e^{-\phi_*/2} \quad \text{at } y = y_* + 0. \tag{42}$$

The problem is therefore reduced to solving the differential equations (40) under the boundary conditions (41), (42). Since there are six boundary conditions on this fourth-order system, we may expect the parameters ϕ_* and y_* to be determined as functions of ε.

Independent solutions of the T-equation (40a) are 1 and erf(δy), where

$$\delta = \left(\frac{\varepsilon}{2}\right)^{1/2}; \tag{43}$$

the boundary conditions (41a), (42a) then show that

$$T = T_f + \frac{Y_f \operatorname{erfc}(\delta y)}{\operatorname{erfc}(d)} \quad \text{with } d = \delta y_*. \tag{44}$$

A particular solution of the h-equation (40b) is now found proportional to $y \exp(-\delta^2 y^2)$, from which we construct the solution

$$h = \frac{-T_b^2 \phi_* \operatorname{erfc}(\delta y)}{\operatorname{erfc}(d)} + \frac{l Y_f [\delta y e^{-\delta^2 y^2} \operatorname{erfc}(d) - d e^{-d^2} \operatorname{erfc}(\delta y)]}{\pi^{1/2} (\operatorname{erfc} d)^2} \tag{45}$$

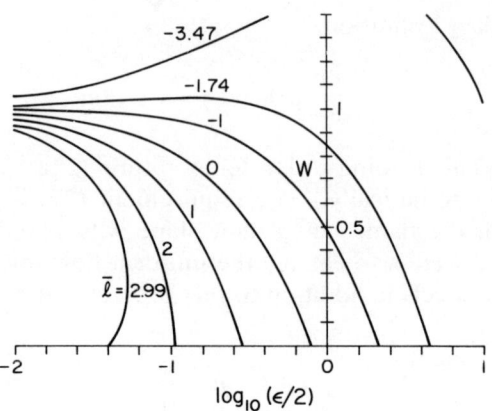

FIG. 4.4. *Variation of flame speed W with straining rate ε in stagnation-point flow.*

satisfying the boundary conditions (41b), (42b). The relation

$$\phi_* = 2\left\{\ln\left[\frac{\pi^{1/2}\,\text{erfc}\,(d)}{2\delta}\right] + d^2\right\} \tag{46}$$

and, finally, the equation

$$\delta = \frac{\pi^{1/2}}{2}\,\text{erfc}\,(d)\exp\left\{d^2 + \left[\frac{de^{-d^2}}{\pi^{1/2}\,\text{erfc}\,(d)} - \frac{1}{2} - d^2\right]\bar{l}\right\} \quad \text{with } \bar{l} = \frac{Y_f l}{2T_b^2}, \tag{47}$$

for the standoff distance y_* as a function of ε, follow from the boundary conditions (42c, d).

Of greatest interest is the flame speed (35a) as a function of stretch (36), and this is plotted in Fig. 4 for various values of $\bar{l}$. As $\varepsilon \to 0$, W tends to the value 1 (that for an adiabatic plane flame). As ε increases from zero, W initially decreases for $\bar{l} > -1$, but increases for $\bar{l} < -1$, in agreement with SVFs for $\mathscr{L} \geq 1$. This behavior is described by the formula

$$W = 1 - \varepsilon(1 + \bar{l}) + O(\varepsilon^2), \tag{48}$$

which can be shown to hold quite generally for flows with small strain. (The formula has implications for stability; see § 5.3.) Further increase in ε leads to two possibilities: for $\bar{l} < 2$, the flame sheet eventually moves to the wall and is extinguished; for $\bar{l} > 2$, extinction takes place in the interior of the flow. This dichotomy is observed in experiments.

In short, stretch usually decelerates the flame and always extinguishes it. Acceleration will occur for weak stretch if the Lewis number is sufficiently far below 1.

Success in treating stagnation-point flow is due to the reduction from partial to (tractable) ordinary differential equations that is effected by the velocity field (33); changing the boundary conditions makes no difference, provided they are independent of x. For example, Buckmaster and Mikolaitis have replaced the wall by an inert counterflow at a temperature close to T_b, and Daneshyar, Ludford & Mendes-Lopes (1983) have considered loss of heat to the wall. Daneshyar, Ludford, Mendes-Lopes & Tromans (1983) have even taken account of expansion through the flame by modifying the velocity field without losing tractability. Finally, Mikolaitis & Buckmaster (1981) have considered rear stagnation-point flow (i.e. $\varepsilon < 0$); see § 7.6.

LECTURE 5

Stability of the Plane Deflagration Wave

Steady, plane deflagration was introduced in the second lecture; here we shall consider infinitesimal perturbations of it and so examine its stability. We shall find two basic phenomena—the hydrodynamic and Lewis-number effects.

Without the constant-density approximation our task is not easy, because the perturbation equations (though of course linear) have variable coefficients. There are three ways in which this difficulty can be overcome:

(i) If attention is restricted to disturbances whose wavelength is much greater than the thickness of the deflagration, a hydrodynamic description is appropriate (§ 3.1). This eliminates the Lewis-number effect, though it may be readmitted as a perturbation (Pelce & Clavin (1982)).

(ii) If the constant-density approximation is adopted, then the T- and Y-equations (the only ones that have to be solved) have constant coefficients. This eliminates the hydrodynamic effect, leaving the Lewis-number effect. The hydrodynamics can be reintroduced in the context of a weakly nonlinear theory based on appropriately small heat release.

(iii) For one-dimensional disturbances, the governing equations can be reduced to their constant-density form by means of a von Mises transformation, the distance variable being replaced by the particle function. This requires the diffusion coefficients to be proportional to T (rather than constants), by no means an unreasonable assumption physically.

In this lecture we shall consider the limiting cases (i) and (ii), thereby isolating the hydrodynamic and Lewis-number effects.

1. Darrieus–Landau instability. The treatment of flame stability from a hydrodynamic viewpoint is due, independently, to Darrieus and Landau. The work of Darrieus, a French aeronautical engineer well known for his invention of the vertical-axis windmill (Darrieus rotor), is often cited as part of the proceedings of the 1946 International Congress of Applied Mechanics, but these were never published. Copies of a 1938 typescript are in the possession of several members of the combustion community.

As we saw in § 3.1, large-scale disturbances of a plane flame are described by Euler's equations on either side of a temperature discontinuity, viz.

$$\nabla \cdot \mathbf{v} = 0, \qquad \rho \frac{D\mathbf{v}}{Dt} = -\nabla p - \rho g \mathbf{i}. \tag{1}$$

Bars have been dropped, so it should be remembered that the scales are much bigger than the diffusion time and length. Note that a gravity term has been added, corresponding (when g is positive) to the burnt gas in $x > 0$ lying above the fresh mixture in $x<0$. The jump conditions (3.9), (3.10) apply across the discontinuity, whose nominal position is $x=0$.

The undisturbed flow, found as a solution of equations (1) satisfying the jump conditions (3.9), (3.10), is

$$\rho_0 = \begin{cases} 1, \\ \frac{1}{\sigma}, \end{cases} \quad \mathbf{v}_0 = \begin{cases} \mathbf{i}, \\ \sigma \mathbf{i}, \end{cases} \quad p_0 = \begin{cases} -gx \\ (1-\sigma) - \frac{gx}{\sigma} \end{cases} \quad \text{for } x \lessgtr 0, \tag{2}$$

where σ is the expansion ratio (3.7). The deformation of the discontinuity is represented by

$$x = F_1(y, t) \tag{3}$$

if the (small) disturbance parameter is absorbed by F_1 (see Fig. 1); we consider perturbations for which

$$F_1 = A \exp(iky + \alpha t) \quad \text{with } k > 0. \tag{4}$$

Restriction to two dimensions, implied by this and

$$\mathbf{v} = (u, v), \tag{5}$$

involves no loss of generality. The goal is to determine the growth parameter α as a function of k, the prescribed wave number of the disturbance. The corresponding perturbations of the flow field are governed by

$$\rho_1 = 0, \quad \nabla \cdot \mathbf{v}_1 = 0, \quad \rho_0 \frac{\partial \mathbf{v}_1}{\partial t} + \rho_0 u_0 \frac{\partial \mathbf{v}_1}{\partial x} = -\nabla p_1, \tag{6}$$

and the problem is complete once we have found the jump conditions satisfied by these perturbations.

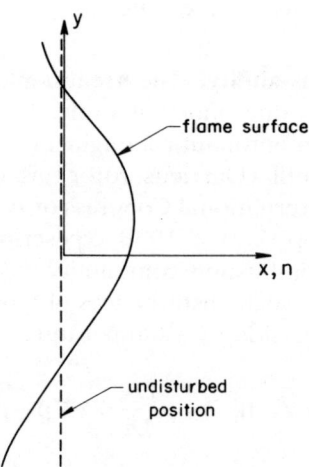

FIG. 5.1. *Notation for stability analysis of plane flame as hydrodynamic discontinuity* (x) *and as NEF* (n). *Flame surface is given by* $F_1(y, t)$.

Since only terms that are linear in disturbance quantities are retained, the unit vectors in the directions normal and tangential to the discontinuity are $(1, -F_{1y})$ and $(F_{1y}, 1)$ so that

$$v_n = u_0 + u_1, \qquad v_\perp = u_0 F_{1y} + v_1. \tag{7}$$

The jump conditions (3.9), (3.10) now give

$$u_{1b} = F_{1t}, \quad v_{1b} - v_{1f} = (1-\sigma)F_{1y}, \quad p_{1b} - p_{1f} = (\sigma^{-1} - 1)gF_1 \tag{8}$$

when we note that

$$V = -F_{1t}. \tag{9}$$

Here, following Darrieus and Landau, we have taken

$$W = 1, \quad \text{i.e. } u_{1f} = F_{1t}, \tag{10}$$

an assumption that can be justified for SVFs with $\mathscr{L} = 1$ (§ 4.2) and for NEFs (§ 4.4). The conditions (8), (10) may all be applied at the undisturbed location of the discontinuity, i.e. $x = 0$.

Solution of the perturbation equations proceeds separately on the two sides of the discontinuity. In the fresh mixture the flow is irrotational (since there are no upstream disturbances), so that

$$u_1 = kP_f e^{kx}, \quad v_1 = ikP_f e^{kx}, \quad p_1 = -(\alpha + k)P_f e^{kx} \quad \text{for } x < 0, \tag{11}$$

if the factor $\exp(iky + \alpha t)$ is omitted. The amplitude P_f of this potential field is as yet unknown. In the burnt mixture there are also rotational terms, due to the generation of vorticity at the curved flame, so that

$$u_1 = kP_b e^{-kx} + kSe^{-\alpha x/\sigma}, \quad v_1 = -ikP_b e^{-kx} - \left(\frac{i\alpha}{\sigma}\right) Se^{-\alpha x/\sigma},$$

$$p_1 = \left(\frac{\alpha}{\sigma} - k\right) P_b e^{-kx} \quad \text{for } x > 0. \tag{12}$$

In addition to P_b, the amplitude S of the solenoidal field is still to be determined.

The requirements (8), (10) at the discontinuity give four homogeneous equations for A, P_f, P_b, S. These have a nontrivial solution if and only if

$$(\sigma+1)\alpha^2 + 2\sigma k\alpha + (\sigma-1)(g - \sigma k)k = 0, \tag{13}$$

i.e., if

$$\alpha = -\frac{\sigma}{\sigma+1} k \pm \sqrt{\frac{\sigma(\sigma^2 + \sigma - 1)}{(\sigma+1)^2} k^2 - \left(\frac{\sigma-1}{\sigma+1}\right) gk}. \tag{14}$$

In the absence of gravity, i.e. for $g = 0$, the larger root is positive for all k, corresponding to instability for all wavelengths. This is the Darrieus–Landau result, which has been an embarrassment to combustion scientists ever since it was discovered. We shall now see how gravity modifies the unacceptable conclusion that plane flames are unstable.

When $g \neq 0$ it is instructive to examine the short and long wavelength limits. As $k \to \infty$,

$$\alpha = \frac{[-\sigma \pm \sqrt{\sigma(\sigma^2 + \sigma - 1)}]k}{\sigma + 1} + \cdots, \tag{15}$$

so that the influence of gravity dies out and short wavelengths are unstable. As $k \to 0$,

$$\alpha = \pm i \sqrt{\frac{(\sigma - 1)gk}{\sigma + 1}} - \frac{\sigma k}{\sigma + 1} + \cdots \tag{16}$$

so that gravity stabilizes long wavelengths. The critical wavenumber separating stable and unstable disturbances is

$$k_c = \frac{g}{\sigma}. \tag{17}$$

Hydrodynamic instability is observed in flames (see, e.g., Sivashinsky (1983)), but it is often absent. There are several ways of reconciling this fact with the present results. The flame may be too small to be treated as a hydrodynamic discontinuity; the wavenumbers that allow the disturbed flame to be so treated may be less than k_c; or there may be other stabilizing influences, such as curvature (§ 4).

2. The Lewis-number effect: SVFs. We now set aside the hydrodynamics and investigate the Lewis-number effect, by adopting the constant-density approximation. The first part of our discussion is concerned with SVFs, which are governed by (4.12) with $\Psi = 0$ if volumetric heat losses are negligible.

Section 3.3 found that such flames are unstable to plane disturbances, even finite ones, for $\mathscr{L} > 1$. Nonplanar disturbances are also unstable, now for $\mathscr{L} < 1$ as well. This can be demonstrated by examining large-scale disturbances, large even compared to θ. To this end, we introduce new variables

$$(\hat{\eta}, \hat{\zeta}, \hat{\tau}) = \delta(\eta, \zeta, \tau) \tag{18}$$

and consider the limit $\delta \to 0$. Then $B = O(\delta)$ in the basic equation (4.12), and $W = 1$ to leading order; to next order, we find

$$W = 1 - \tfrac{1}{2}\delta b \hat{K} \quad \text{with} \quad \hat{K} = S^{-1} \frac{dS}{d\hat{\tau}}, \tag{19}$$

an explicit formula for the effect of weak (superficial) stretch on the flame speed. Clearly such stretch decreases the wave speed only for $\mathscr{L} > 1$ (cf. § 4.3).

The instability of the plane flame for $\mathscr{L} < 1$ follows immediately from this relation. Suppose that, because of some disturbance, corrugations have formed (Fig. 2). The troughs, as viewed from the burnt mixture, experience positive stretch, while the crests are compressed. It follows from (19) with $b < 0$ that the flame speed at the troughs is increased while that at the crests is decreased, so that the amplitude of the corrugations will grow (corresponding to instability).

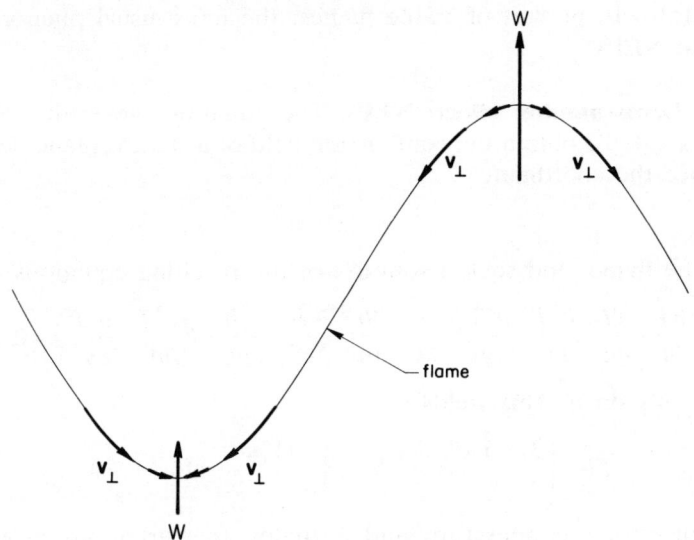

FIG. 5.2. *Instability of SVF for $\mathscr{L}<1$.*

We are dealing with a cellular instability, whose nature is most clearly seen for the NEFs discussed below. Because of the connection (3.27) between the speed and temperature of the flame, the crests are relatively cold and, hence, less luminous than the rest of the flame. For $\mathscr{L}>1$ the effect is reversed, a stable situation.

That this stability conclusion remains valid for disturbances of more moderate scale can also be seen from the basic equation (4.12) of SVFs. Setting

$$W = 1 + W_1, \quad K = \kappa_1 \tag{20}$$

gives

$$b\left(\frac{dW_1}{d\tau} - \kappa_1\right) = 2W_1 \quad \text{with } W_1 = -F_{1\tau}, \quad \kappa_1 = F_{1\eta\eta} \tag{21}$$

if

$$\chi = -\tau + F_1(\eta, \tau) \tag{22}$$

is the location of the disturbed flame sheet. Growth of the mode proportional to $\exp(\alpha\tau + ik\eta)$ is therefore determined by

$$\alpha^2 - 2b^{-1}\alpha - k^2 = 0, \quad \text{i.e. } \alpha = b^{-1} \pm \sqrt{b^{-2} + k^2}. \tag{23}$$

For $k \neq 0$, one of the roots is positive whatever the value of b (i.e. $\mathscr{L}$), so that there is instability for all nonplanar disturbances when $\mathscr{L} \neq 1$. For $k = 0$, there is instability only when b is positive, i.e. for $\mathscr{L} > 1$.

The instability of SVFs restricts their value as a framework within which to discuss flames, although the qualitative insights obtained from such solutions can be very useful (e.g. concerning flame tips, see Buckmaster & Ludford

(1982, p. 159). In pursuit of stable flames, the more usual phenomenon, we now turn to NEFs.

3. The Lewis-number effect: NEFs. The equations governing NEFs were derived in § 4.4. To obtain the combustion field of a steady, plane deflagration, we introduce the coordinate

$$n = x + t \tag{24}$$

based on the flame, and seek a solution of the resulting equations

$$\frac{\partial T}{\partial t} + \frac{\partial T}{\partial n} - \frac{\partial^2 T}{\partial n^2} - \frac{\partial^2 T}{\partial y^2} = \frac{\partial h}{\partial t} + \frac{\partial h}{\partial n} - \frac{\partial^2 h}{\partial n^2} - \frac{\partial^2 h}{\partial y^2} - l\left(\frac{\partial^2 T}{\partial n^2} + \frac{\partial^2 T}{\partial y^2}\right) = 0 \tag{25}$$

depending only on n. This yields

$$T_0 = \begin{cases} T_f + Y_f e^n, \\ T_b, \end{cases} \quad h_0 = \begin{cases} -lY_f n e^n \\ 0 \end{cases} \quad \text{for } n \lessgtr 0 \tag{26}$$

as the undisturbed temperature and enthalpy (perturbation) profiles, since equilibrium must prevail behind the flame.

If the equation of the disturbed flame sheet (see Fig. 1) is

$$n = F_1(y, t), \tag{27}$$

so that the normal derivative in the jump conditions (4.27)–(4.29) becomes $\partial/\partial n - F_{1y} \partial/\partial y$, then these conditions may be written

$$T_1^- = -Y_f F_1, \quad \delta(h_1) = -lY_f F_1, \quad \frac{\partial T_1^-}{\partial n} = -Y_f F_1 + \frac{h_1^+}{l_s},$$

$$\delta\left(\frac{\partial h_1}{\partial n}\right) = \frac{lh_1^+}{l_s} - 2lY_f F_1 \quad \text{with } l_s = \frac{2T_b^2}{Y_f}, \tag{28}$$

all quantities being evaluated at the undisturbed flame sheet $n = 0$. (We have used the relation $\phi_* = -h_1^+/T_b^2 + \cdots$ and superscripts $\pm$ to denote values at $n = 0\pm$.) The problem is to solve (25) subject to these conditions and the requirement that T_1, h_1 die out as $n \to \pm\infty$. Arbitrary initial conditions are, as usual, taken into account by considering perturbations for which

$$F_1 = A \exp(iky + \alpha t). \tag{29}$$

The possibility of stability can be seen most easily for $l = 0$, when the jump conditions

$$\delta(h_1) = \delta\left(\frac{\partial h_1}{\partial n}\right) = 0 \tag{30}$$

ensure that

$$h_1 = 0 \quad \text{for all } n \tag{31}$$

is the appropriate solution of the differential equation (25b). The remaining problem is to solve the T_1-equation for $n < 0$ alone, subject to the boundary conditions

$$T_1 = \frac{\partial T_1}{\partial n} = -Y_f A \quad \text{at } n=0; \tag{32}$$

the factor $\exp(iky+\alpha t)$ has been omitted. Now, the solutions of the perturbation equation are

$$e^{(1\pm\kappa)n/2} \quad \text{with } \kappa = \sqrt{1+4\alpha+4k^2} \quad \left(-\frac{\pi}{2} < \arg \kappa \le \frac{\pi}{2}\right) \tag{33}$$

where, if attention is restricted to unstable modes $\operatorname{Re}(\alpha)>0$, we have

$$\operatorname{Re}(\kappa) > 1. \tag{34}$$

It follows that the appropriate solution is

$$T_1 = Be^{(1+\kappa)n/2} \quad \text{for } n<0, \tag{35}$$

and then the boundary condition (32a) requires

$$B = \tfrac{1}{2}(1+\kappa)B, \quad \text{i.e. } \kappa = 1 \text{ or } \alpha = -k^2. \tag{36}$$

This contradicts our assumption that the mode is unstable, so we must conclude that there are no unstable modes: when $\mathscr{L}=1$ the flame is stable for all wavenumbers k.

From the results obtained in § 2 for slowly varying disturbances, we may expect that, as l becomes large (positively or negatively), long wavelength disturbances (k small) will become unstable. To see how this happens, we now consider the jump conditions (28) for $l \ne 0$. Solution of the perturbation equations proceeds separately on the two sides of the flame sheet. In front we find

$$T_1 = Be^{(1+\kappa)n/2}, \quad h_1 = \left\{C + l\kappa^{-1}\left[k^2 - \frac{(1+\kappa)^2}{4}\right]Bn\right\}e^{(1+\kappa)n/2} \quad \text{for } n<0, \tag{37}$$

where κ has the definition (33b) and again the factor $\exp(iky+\alpha t)$ has been omitted. Behind we have

$$T_1 = 0, \quad h_1 = De^{(1-\kappa)n/2} \quad \text{for } n>0. \tag{38}$$

The expressions are valid only for $\operatorname{Re}(1+\kappa)>0$ and $\operatorname{Re}(1-\kappa)<0$, i.e. when the condition (34) is satisfied, as it is for unstable modes. The jump conditions (28) now yield a homogeneous system for the coefficients A, B, C, D; a nontrivial solution exists only if

$$2\kappa^2(1-\kappa) + \bar{l}[(1-\kappa)^2 - 4k^2] = 0 \quad \text{with } \bar{l} = \frac{l}{l_s}, \tag{39}$$

a result due to Sivashinsky.

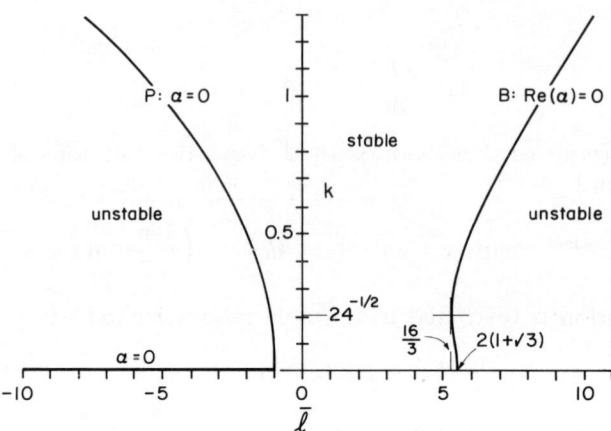

FIG. 5.3. *Linear stability regions for plane NEFs. The boundaries are* P: $4k^2 = -(\bar{l}+1)$; B: $\bar{l} = 2(1+8k^2)[1+\sqrt{3(1+8k^2)}]/(1+12k^2)$.

This dispersion relation should be viewed as determining, for each $\bar{l}$, the growth parameter α of any unstable disturbance mode of wave number k. The stability boundaries in the $\bar{l}$, k-plane therefore correspond to Re $(\alpha) = 0$; they are shown in Fig. 3. At any wave number there is a finite band of Lewis numbers, always including $\mathscr{L} = 1$, for which the flame is stable and outside of which it is unstable. The left boundary, on which Im (α) vanishes also, is a creation of nonplanar disturbances: the dispersion relation (39) does not yield an α with positive real part for $\bar{l} < -1$ and $k = 0$. In fact, the unstable mode becomes neutral as this part of the $\bar{l}$-axis is approached. On the other hand, the instability predicted by the right boundary, where Im $(\alpha) \neq 0$, does survive a planar treatment: the dispersion relation (39) with $k = 0$ does yield an α with positive real part for $\bar{l} > 2(1+\sqrt{3})$.

The stability boundary in the neighborhood of the point $\bar{l} = -1$, $k = 0$ can be explained by flame stretch, as was the destabilizing effect of very long wavelength disturbances for $\mathscr{L} < 1$ in § 2. We need only note the relation (4.48) between weak stretch and flame speed: for $\bar{l} < -1$, positive (negative) stretch increases (decreases) the wave speed, tending to increase the amplitude of any long-wavelength disturbance (cf. Fig. 2); for $\bar{l} > -1$, amplitudes are decreased.

There are convincing reasons for believing that the left boundary is associated with cellular flames, a common laboratory phenomenon: it corresponds to $\mathscr{L} < 1$, the Lewis numbers for which such flames are seen; the crests of disturbed flames are darker than the troughs (just as for SVFs with $\mathscr{L} < 1$), this being a characteristic of cellular flames; and, for $\bar{l} < -1$, all modes with $k < \frac{1}{2}\sqrt{-(1+\bar{l})}$ are unstable, suggesting that the outcome of the instability will not be monochromatic, another characteristic of cellular flames. In Lecture 6 we shall give a nonlinear theory that arises naturally from the present linear analysis and reinforces this conviction.

The right stability boundary is associated with pulsations or traveling waves; it is relatively inaccessible because $\mathscr{L}$ is rarely much bigger than 1. However, similar phenomena may be expected in so-called thermites, for which $\mathscr{L} = \infty$. Otherwise, special means must be devised to make the boundary more accessible. These matters form the subject of Lecture 7.

4. The role of curvature. The plane premixed flame whose stability has been discussed so far is an idealization seldom approximated, since in practice the flame is usually curved. Even under circumstances designed to nurture a plane state, imperfections can thwart the best efforts and give a curvature, albeit weak, to the flame. In this section we shall investigate certain slightly curved flames, those amenable to the SVF analysis in §2 and those associated with the left stability boundary for NEFs in §3.

Consider the cylindrical source/sink flow

$$u = \pm \frac{R}{r}, \qquad v = 0, \tag{40}$$

where θr is radial distance and u, v are now polar components. The undisturbed flame is then circular, with θ-multiplied curvature

$$\kappa = \mp R^{-1} \tag{41}$$

determined by the requirement that the mass flux through the unstretched flame be 1. The expressions (20) are replaced by

$$W = 1 + W_1, \qquad K = \nabla_\perp \cdot \mathbf{v}_{1\perp} + \kappa_1 + \kappa W_1, \tag{42}$$

where $\mathbf{v}_{1\perp}$ is due to the displacement of the flame in the source/sink flow, which (for the constant-density approximation) is undisturbed. Substitution in the basic equation (4.12) gives

$$b\left(\frac{dW_1}{d\tau} - \nabla_\perp \cdot \mathbf{v}_{1\perp} - \kappa_1 - \kappa W_1\right) = 2W_1 \tag{43}$$

with

$$W_1 = -(R^{-1}F_1 \pm F_{1\tau}), \quad \nabla_\perp \cdot \mathbf{v}_{1\perp} = \mp 2R^{-2}F_1, \quad \kappa_1 = \pm R^{-2}(F_1 + F_{1\phi\phi}) \tag{44}$$

if

$$r = R + F_1(\phi, \tau) \tag{45}$$

is the location of the disturbed flame sheet. The behavior of the mode proportional to $\exp(\alpha\tau + in\phi)$ is therefore given by

$$\alpha^2 - 2(b^{-1} \mp R^{-1})\alpha - R^{-1}(\pm 2b^{-1} + n^2 R^{-1}) = 0. \tag{46}$$

To discuss this result we introduce the wavenumber

$$k = R^{-1}n \tag{47}$$

and note that the equation (23a) is recovered as $R \to \infty$. There will be instability for those values of k, R that make one of the (always real) roots of

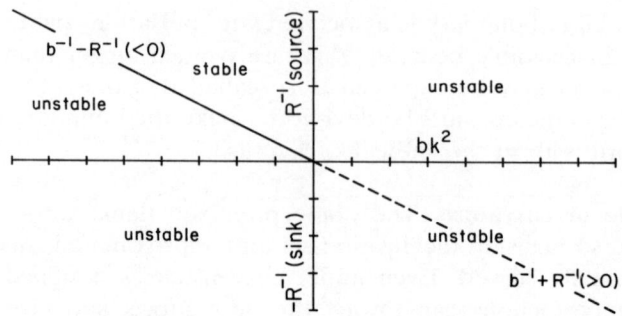

FIG. 5.4. *Effect of curvature on SVF stability.*

equation (46) positive. Such values are bounded by ones for which $\alpha = 0$, but the reverse is not true. Figure 4 shows the line on which one root vanishes and one half of the corresponding value of the other root; the full part of the line is therefore a stability boundary but the dashed part is not. Results for $R \to \infty$, given in the discussion of (23), then show that the stability regions are as labelled. We conclude that curvature is never stabilizing for a sink flow, but that it is for a source flow when

$$\mathscr{L} < 1, \quad k < \sqrt{\frac{-2}{bR}}. \tag{48}$$

We turn now to NEFs and the neighborhood of $\bar{l} = -1$, $k = 0$ in Fig. 3, where the dispersion relation (39) reduces to

$$\alpha = -(1 + \bar{l})k^2 - 4k^4; \tag{49}$$

to have a balanced equation we must require

$$k^2 = O(1 + \bar{l}), \quad \text{so that } \alpha = O((1 + \bar{l})^2). \tag{50}$$

To see how this relation is modified by curvature, we consider once more a circular flame $r = R$ sustained by a source/sink flow. A derivation of the new dispersion relation will be given in § 6.2, where it is needed for a nonlinear analysis. Here we shall only note that the connection between wave speed and stretch, represented by the generalization (49) of the relation (4.48), is preserved; but that recalculations of the stretch, represented by the term in k^2, and of the wave speed, represented by the term α, are needed because of the nonuniformity of the velocity field.

The modified dispersion relation is

$$\alpha = -(1 + \bar{l})k^2 - 4k^4 \mp R^{-1}, \tag{51}$$

where k is now limited to the values (47); a balanced equation requires

$$R = O((1 + \bar{l})^{-2}) \tag{52}$$

in addition to the earlier restriction (50a). For $\bar{l} < -1$ and no curvature, there is

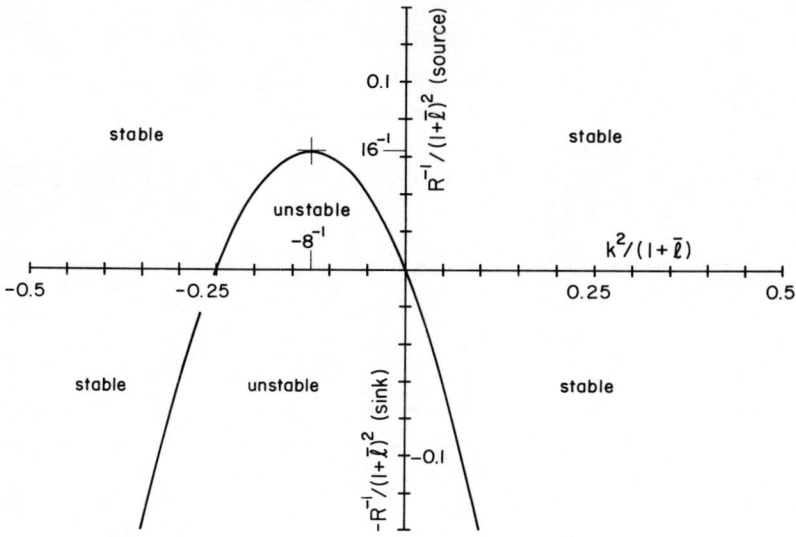

FIG. 5.5. *Effect of curvature on NEF stability.*

a band of wavenumbers

$$0 < k < \tfrac{1}{2}\sqrt{-(1+\bar{l})} \tag{53}$$

for which α is positive, corresponding to instability (cf. Fig. 3). Source flow, i.e. convexity of the flame sheet towards the burnt mixture, narrows the band (Fig. 5) and, for sufficiently high curvature

$$R^{-1} > \frac{(1+\bar{l})^2}{16}, \tag{54}$$

eliminates it. Sink flow, on the other hand, widens the band indefinitely. For $\bar{l} > -1$ and no curvature, α is negative for all k, corresponding to stability. Source flow does not change this conclusion, but sink flow produces instability for wavenumbers below an ever-increasing upper limit (Fig. 5).

For both the SVF and the NEF, convexity towards the burnt mixture is found to be stabilizing, concavity destabilizing.

LECTURE 6

Cellular Flames

We shall now examine the left stability boundary that was uncovered in Lecture 5 in our discussion of NEFs (Fig. 5.3). The boundary is associated with instabilities leading to cellular flames, i.e. flames whose surfaces are broken up into distinct luminous regions (cells) separated by dark lines. Each line is a ridge of high curvature, convex towards the burnt gas. For a nominally flat flame these cells are very unsteady, growing and subdividing in a chaotic fashion; but curvature, for example, can make them stationary.

The most striking manifestation of cellular instability is the polyhedral flame, into which the conical flame on a Bunsen burner can suddenly transform. The conical surface splits into triangular cells forming a polyhedron; for a five-sided flame the appearance, from above, is much like that of the Chrysler emblem (Fig. 1). The dark wedges between the white triangular cells correspond to sharp ridges; the dark central region corresponds to a tip with strong curvature. Figure 2 gives a sketch of a five-sided flame, derived from a photograph in Smith & Pickering (1929). Polyhedral flames are often stationary, but can spin rapidly about the vertical axis, making several revolutions per second.

We shall discuss chaotic and stationary cellular flames, including polyhedral flames, in the framework of the weakly nonlinear theory pioneered by Sivashinsky. The constant-density approximation will be used throughout, although perturbations of it will be admitted in two places.

1. Chaotic cellular structure. The nonlinearity associated with the left stability boundary will be weakest in the neighborhood of

$$\bar{l} = -1, \quad k = 0, \tag{1}$$

a possible bifurcation point; accordingly we focus our attention there by taking

$$\bar{l} + 1 = O(\varepsilon), \quad k = O(\sqrt{\varepsilon}), \quad \alpha = O(\varepsilon^2), \tag{2}$$

where ε is a small positive parameter that will be found to represent the amplitude of the disturbance. The relative ordering of $\bar{l}+1$ and k is suggested by the parabolic shape of the stability boundary, while the order of α follows from the limiting form (5.48) of the dispersion relation (5.39) as $\varepsilon \to 0$. This determines the growth rate of the most important Fourier components (the unstable ones) of the disturbance when $\bar{l}+1$ is small. In terms of any scalar F that represents the disturbance field, the dispersion relation is equivalent to

$$F_t + 4F_{yyyy} - (1+\bar{l})F_{yy} = 0. \tag{3}$$

For $\bar{l}+1<0$, this equation predicts unbounded growth. Bifurcation (with weakly nonlinear description) is possible if nonlinear effects, not yet taken into

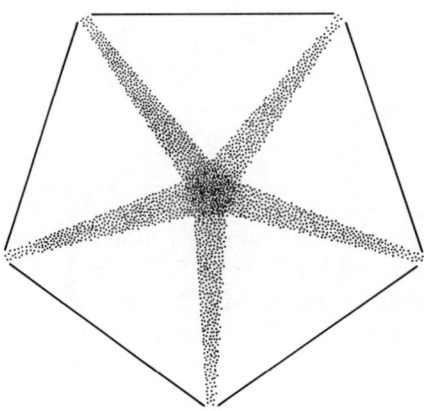

FIG. 6.1. *Chrysler emblem.*

account, limit this growth. We shall first give a heuristic argument to determine these effects and then substantiate the result by formal analysis. The argument consists in recognizing that (3) is actually a formula for the wave speed, and modifying it appropriately. In this connection, suppose that F determines the location of the flame sheet as

$$x = -t + \varepsilon F; \tag{4}$$

then the speed of the sheet is

$$W = 1 + \varepsilon W_1 + \cdots \quad \text{with } W_1 = -F_t, \tag{5}$$

and (3) becomes

$$W_1 = 4F_{yyyy} - (1 + \bar{l})F_{yy}. \tag{6}$$

This formula determines the deviations of the flame speed from its adiabatic

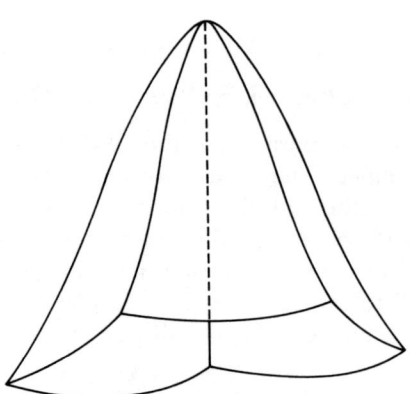

FIG. 6.2. *Five-sided polyhedral flame.*

value of 1 due to the reaction-diffusion effects that are triggered by the distortion of the flame front.

Now, (5b) is mere kinematics, valid in a linear theory only. The exact relation between flame speed and displacement is

$$W = \frac{1 - \varepsilon F_t}{\sqrt{1 + \varepsilon^2 F_y^2}} = 1 - \varepsilon F_t - \tfrac{1}{2}\varepsilon^2 F_y^2 + \cdots \tag{7}$$

and, for disturbances with wave numbers of the magnitude (2b), the nonlinear terms $\tfrac{1}{2}\varepsilon^2 F_y^2$ is comparable to the linear term εF_t. This suggests that the nonlinear generalization

$$W_1 = -F_t - \tfrac{1}{2}\varepsilon F_y^2 \tag{8}$$

should be used in the formula (6) and, when ε is purged from the resulting equation by writing

$$\bar{l} + 1 = -\varepsilon, \quad \eta = \sqrt{\varepsilon} y, \quad \tau = \varepsilon^2 t \tag{9}$$

(in accordance with the ordering (2)), we find

$$F_\tau + \tfrac{1}{2}F_\eta^2 + 4F_{\eta\eta\eta\eta} + F_{\eta\eta} = 0. \tag{10}$$

Note that this equation holds for $\bar{l} < -1$.

Substantiation of this result requires a systematic asymptotic development in which x is replaced by the coordinate

$$n = x + t - \varepsilon F(\eta, \tau) \tag{11}$$

in the governing equations (4.24), (4.25); thus,

$$\frac{\partial}{\partial x} = \frac{\partial}{\partial n}, \quad \frac{\partial}{\partial y} = -\varepsilon^{3/2} F_\eta \frac{\partial}{\partial n} + \varepsilon^{1/2} \frac{\partial}{\partial \eta}, \quad \frac{\partial}{\partial t} = (1 - \varepsilon^3 F_\tau) \frac{\partial}{\partial n} + \varepsilon^2 \frac{\partial}{\partial \tau} \tag{12}$$

when y, t are replaced by η, τ. The normal derivative, required for the jump conditions (4.27)–(4.29), is

$$(1 + \tfrac{1}{2}\varepsilon^3 F_\eta^2) \frac{\partial}{\partial n} - \varepsilon^3 F_\eta \frac{\partial}{\partial \eta} \tag{13}$$

to sufficient accuracy. Perturbation expansions in ε are now introduced for T, h and F, leading to a sequence of linear problems for the T- and h-coefficients as functions of n, η and τ. These are to be solved under the requirements: $T_1 = T_2 = T_3 = \cdots = 0$ for $n > 0$; conditions as $n \to -\infty$ are undisturbed; and exponential growth as $n \to +\infty$ is disallowed. The problems are overdetermined, but only at the fourth (for T_3, h_3) is a solvability condition required, namely (10) for the leading term in F.

For two-dimensional disturbances of the flame sheet, the basic equation is

$$F_\tau + \tfrac{1}{2}(\nabla F)^2 + 4\nabla^4 F + \nabla^2 F = 0. \tag{14}$$

Discussion for both one- and two-dimensional disturbances has been limited to numerical computations. The solutions obtained display chaotic variations in a

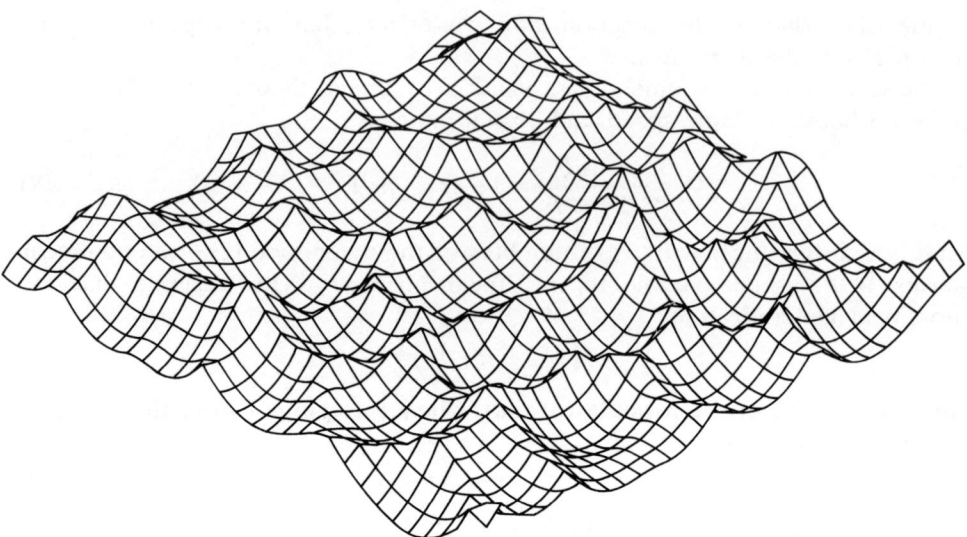

FIG. 6.3. *Numerical calculation of cellular flame.* (Courtesy G. I. Sivashinsky.)

cellular structure, resembling the behavior of actual flames. Figure 3 clearly shows the ridges that separate the individual cells.

2. Effect of curvature. Equation (10) is a balance of small terms; it may be modified to account for any additional physical process whose effect is also small. Hydrodynamic effects can be incorporated, for example, if the density change across the flame is appropriately small (because of small heat release), and this provides important insight into the role of Darrieus–Landau instability in actual flames. Equation (10) is replaced by

$$F_\tau + \tfrac{1}{2} F_\eta^2 + 4 F_{\eta\eta\eta\eta} + F_{\eta\eta} + \gamma \!\!\not{\!\!\int}_{-\infty}^{\infty} \frac{F_\eta(\bar\eta, \tau)}{\bar\eta - \eta} \, d\bar\eta = 0 \quad \text{with } \gamma = \frac{\sigma - 1}{2\pi \varepsilon^{3/2}}, \qquad (15)$$

the requirement on the heat release being $\gamma = O(1)$: the expansion ratio σ can only differ from 1 by $O(\varepsilon^{3/2})$. Numerical integration shows that the new (integral) term is destabilizing; an even finer structure is superimposed on the chaotic cellular pattern obtained without it. We shall not consider the topic further, since Sivashinsky (1983) has recently discussed it in detail. Instead, we shall examine a much simpler effect, that of curvature.

Consider a line source of mixture supporting a stationary cylindrical flame; the radial speed of the efflux is taken to be

$$u = \frac{R}{r}, \qquad (16)$$

where R is an assignable constant. The corresponding solution of equations (4.24), (4.25), the jump conditions (4.27)–(4.29), and the boundary conditions at the origin and infinity show that the flame is located at $r = R$ and that

$$T = \begin{cases} T_f + Y_f\left(\dfrac{r}{R}\right)^R, \\ T_f + Y_f, \end{cases} \quad h = \begin{cases} \dfrac{lY_f r^R \ln(R/r)}{R^{R-1}} \\ 0 \end{cases} \quad \text{for } r \leq R. \tag{17}$$

Note that the speed (16) is 1 at the flame location, an unexpected result. Apparently the effect of flame curvature on its speed, normally significant, is here cancelled by the effect of flow divergence.

When R is $O(\varepsilon^{-2})$ the curvature is $O(\varepsilon^2)$, as for the disturbed plane flame described by (10). We would therefore expect such large flames, if disturbed to the same extent, to be described by a modification of that equation; the additional terms will be due to flow divergence and undisturbed curvature. The modified equation can be derived by formal expansion, as was the original; but such an exercise, although reassuring, is hardly illuminating. We shall instead give a heuristic derivation that emphasizes the physics.

The general effect of large-scale (and therefore weak) stretch on flame speed was identified in § 4.5; we may write

$$W = 1 - (1 + \bar{l})K + \cdots, \tag{18}$$

where K is the Karlovitz stretch (4.3). This effect is the origin of the second term on the right side of (6); the first, corresponding to what is normally a small correction to the result (18), must be retained when $\bar{l}$ is close to -1. In the present context, where flow divergence generates a stretch R^{-1} of order ε^2 in the undisturbed plane flame, (6) has to be replaced by

$$W_1 = 4F_{yyyy} + \varepsilon F_{yy} + R^{-1}. \tag{19}$$

Note that F still represents disturbance of a plane flame (i.e., $x = R + \varepsilon F$) so that the description is valid only up to $O(\varepsilon^{-1/2})$ values of y.

The kinematic expression (8) for the wave speed is also modified, because the flame sheet is moving in a nonuniform velocity field. We find

$$W_1 = -F_t - \frac{yF_y}{R} - \frac{F}{R} - \frac{y^2}{\varepsilon R^2} - \tfrac{1}{2}\varepsilon F_y^2 \tag{20}$$

to sufficient accuracy, so that combination with the result (19) now yields

$$F_t + \frac{\varepsilon}{2}F_y^2 + 4F_{yyyy} + \varepsilon F_{yy} + \frac{F}{R} + \frac{yF_y}{R} = -\frac{y^2}{\varepsilon R^2} - \frac{1}{R}. \tag{21}$$

This has the stationary solution

$$F = F_0 \equiv -\frac{y^2}{2\varepsilon R} \tag{22}$$

corresponding to the undisturbed circular flame. Replacing F by $F + F_0$, so that F now represents disturbance of the circular flame sheet, and using the scaled variables (9) yield

$$F_\tau + \frac{1}{2}F_\eta^2 + 4F_{\eta\eta\eta\eta} + F_{\eta\eta} + \gamma F = 0 \quad \text{with } \gamma = \frac{1}{\varepsilon^2 R}. \tag{23}$$

Comparison with (10) shows that the only new term is γF.

The linearized form of (23), with F set proportional to $\exp(\alpha t + iky)$, was considered in § 5.4. Figure 5.5 shows that curvature is a stabilizing influence, but that instability occurs for

$$\gamma < \gamma_c = \tfrac{1}{16}, \tag{24}$$

corresponding to a supercritical bifurcation with wavenumber

$$k_c = \sqrt{\frac{\varepsilon}{8}}. \tag{25}$$

To show this we write

$$\gamma = \gamma_c - \delta^2, \quad F = \delta f, \quad \tau = \frac{\tilde{\tau}}{\delta^2} \tag{26}$$

and expand f in a power series in δ. In the usual way, we find that the leading term is of the form

$$f_0 = A(\tilde{\tau})e^{ik_c y} + \bar{A}(\tilde{\tau})e^{-ik_c y}, \tag{27}$$

where

$$\frac{dA}{d\tilde{\tau}} = A - \frac{\bar{A}A^2}{36} \tag{28}$$

if there is to be no secular term in the second perturbation of f. The equation describes the evolution in (slow) time $\tilde{\tau}$ of the amplitude from some initial value to the final value $|A| = 6$.

An examination of the first perturbation of f reveals that the crests of the final stationary solution, as viewed from the burnt gas, are sharper than the troughs. Moreover, the flame temperature is diminished at the crests (as for the cellular flames discussed in § 5.3), so that they are darker than the rest of the flame sheet. Sharp, dark crests are a universal feature of cellular flames as observed.

3. Flames near a stagnation point. Equation (23) is only one of a class of evolution equations that describe cellular flames in a variety of circumstances. An unusual example corresponds to a flame located in weak stagnation-point flow

$$\mathbf{v} = \beta(-x, y) \quad \text{with } \beta = O(\varepsilon^2). \tag{29}$$

Two changes have been made in the notation of § 4.5: the x, y-axes have been rotated, so that the wall is now $x = 0$ and the undisturbed flame at $x = x_* < 0$ (cf. Fig. 4.3), to conform with notation already established in this lecture; and the strain rate is now β, since the ε used there has been conscripted as a small parameter here. As for the cylindrical flame, we shall be content with a heuristic derivation.

The undisturbed flame experiences a stretch β so that, if it is displaced by an amount εF, its speed is

$$W_1 = 4F_{yyyy} + \varepsilon F_{yy} + \beta \qquad (30)$$

to sufficient accuracy (cf. (19)). The kinematic expression for the wave speed, corresponding to the result (20), is

$$W_1 = -\frac{\beta x_* + 1}{\varepsilon} - F_t - \beta y F_y - \beta F + \tfrac{1}{2}\varepsilon \beta x_* F_y^2. \qquad (31)$$

Equating these two expressions for W_1 gives a formula for x_*, namely

$$x_* = \frac{1}{\varepsilon^2 \gamma} + \varepsilon \quad \text{with } \gamma = \frac{\beta}{\varepsilon^2}, \qquad (32)$$

and an evolution equation for F:

$$F_\tau + \tfrac{1}{2}F_\eta^2 + 4F_{\eta\eta\eta\eta} + F_{\eta\eta} + \gamma(\eta F)_\eta = 0. \qquad (33)$$

Comparison with (10) shows that the only new term is $\gamma(\eta F)_\eta$.

The generalization

$$F_\tau + \tfrac{1}{2}(\nabla F)^2 + 4\nabla^4 F + \nabla^2 F + \gamma(\eta F)_\eta = 0 \qquad (34)$$

accounts for disturbances that vary in the z-direction also. If we now consider disturbances independent of η, this equation reduces to the earlier one (23) with η replaced by $\zeta = \sqrt{\varepsilon}\, z$. Setting F proportional to $\exp(\alpha t + ikz)$ and linearizing therefore leads to the dispersion relation (5.51) and hence to Fig. 5.5. The bifurcation analysis starting with the transformation (26) is applicable, so that for values of γ slightly smaller than $\tfrac{1}{16}$ there will be a stationary structure characterized by dark ridges pointing towards the burnt gas.

This phenomenon has apparently been known for many years. For upward propagation through sufficiently lean hydrogen-air mixtures in a standard flammability tube, the flame cap is divided into a number of bright strips or ribbons, separated by dark lines (Fig. 4); it seems probable that this is the axisymmetric analog of the nominally plane flame considered here. The straining flow is generated by the gravity-induced convection of the light burnt gas behind the flame (see, for example, Buckmaster & Mikolaitis (1982)).

A different type of disturbance (which can be combined with the previous one without, however, adding to the discussion) corresponds to

$$F = A(\tau)\exp(ik\eta e^{-\gamma\tau}). \qquad (35)$$

Substitution into the linearized version of the evolution equation (33) gives

$$\frac{dA}{d\tau} = A(k^2 e^{-2\gamma\tau} - 4k^4 e^{-4\gamma\tau} - \gamma), \qquad (36)$$

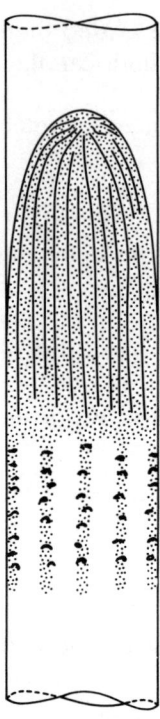

FIG. 6.4. *Flame in a standard flammability tube.*

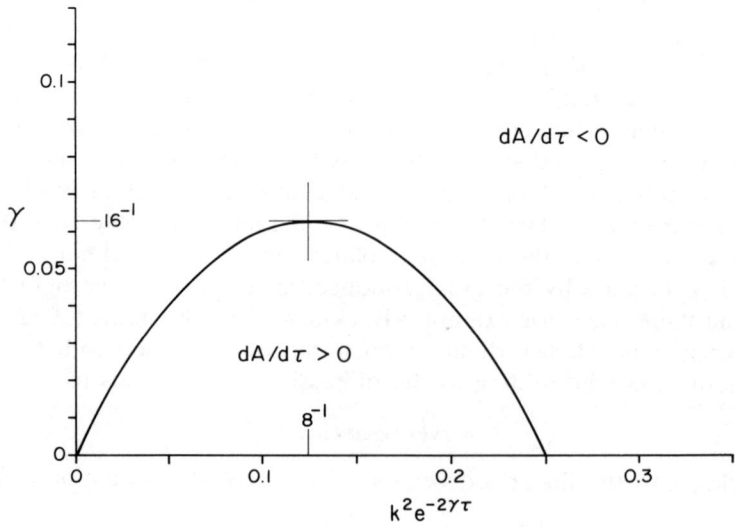

FIG. 6.5. *Curve in $(k^2 e^{-2\gamma\tau}, \gamma)$-plane determining sign of right side of* (36).

which has the solution

$$A = A_0 \exp\left[\frac{(1-e^{-2\gamma\tau})k^2}{2\gamma} + (e^{-4\gamma\tau} - 1)k^4 - \gamma\tau\right]. \tag{37}$$

Figure 5 shows that any disturbance eventually has a decreasing amplitude, although for a time the amplitude increases if γ is less than $\frac{1}{16}$. In the limit $\tau \to \infty$, the solution (37) tends to zero. We conclude that the flame is stable to this type of disturbance.

The chaotic cellular instability found experimentally for weak straining suggests that all disturbances should grow. Moreover, in the limit $\gamma \to 0$ the theoretical results in Lecture 5 predict instability for all (small) wavenumbers. These facts are at variance with the conclusion above, which prompted Sivashinsky, Law & Joulin (1982) to provide the following explanation.

If the nonlinear term is retained in (33), harmonics are continually generated and these may grow during part of their lifetimes, providing a mechanism for sustaining the overall growth of the disturbance into instability. Numerical computations confirm this notion and suggest that the necessary values of γ are significantly smaller than $\frac{1}{16}$. There may be implications for the hydrogen flame of Fig. 4. Away from the nose the rate of strain will be diminished; and may be small enough in the skirt to permit instabilities in the direction of the flow. That is, the ribbon instability may become a cellular instability. Interestingly enough, the tails of the ribbons are often seen to break up into small balls of flame (Fig. 4).

4. Polyhedral flames. In the 90 years since Smithells first observed these flames, they have become established as a familiar laboratory curiosity. They are associated with tube burners, but analogs can be created with different geometries: Markstein (1964, p. 81) designed a slot burner on which he observed a cellular flame behaving essentially like an unwrapped (linear) polyhedral flame. In particular, the cells could be made to travel rapidly from one end of the slot to the other, just as the polyhedral flame can be made to spin. Markstein's photographs of the flame showed that the travelling corrugations are saw-toothed in shape (Fig. 6).

It is the propagation that distinguishes polyhedral flames from other types of cellular instability, so that will be the focus of our discussion. Since the left

BURNT GAS

FRESH GAS

FIG. 6.6. *Analog of spinning polyhedral flame for slot burner.*

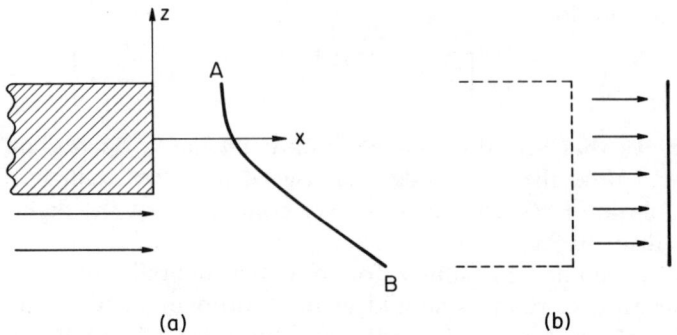

FIG. 6.7. (a) *Behavior of tube flame near rim of burner*; (b) *plane model of* (a) *used to describe polyhedral flames*.

stability boundary is not associated in any obvious way with propagating disturbances (unlike the right stability boundary), the challenge is to uncover a mechanism for such behavior.

One of the difficulties with polyhedral flames is that the undisturbed flame is conical, i.e. nonplanar. Following Buckmaster (1984), we shall overcome this obstacle by adopting a nominally planar model. Consider the portion AB of the burner flame that is located near the rim (Fig. 7a). The flame speed varies from a small value (perhaps zero) at A, to a value comparable to the adiabatic flame speed at B. This portion is modeled by a plane flame with some intermediate speed and standoff distance (Fig. 7b). Perturbations of the planar configuration are permitted in the y-direction, which is perpendicular to the page and parallel to the rim. Corrugations that arise in this way are associated with the corrugations along the entire length of the nominally conical-shaped flame whose base is being modeled.

In the context of the weakly nonlinear theory, perturbations are governed by (10), provided a term is added to account for the presence of the rim. The rim is a heat sink that anchors the flame in a simple fashion: an increase (decrease) in the standoff distance reduces (increases) the heat loss to the rim, thereby increasing (decreasing) the flame speed, a restorative mechanism. If εF is now the perturbation of the standoff distance, this effect can be represented by the modification

$$W_1 = 4F_{yyyy} - (1+\bar{l})F_{yy} + qF \tag{38}$$

of (6). Here q is a positive constant of order ε^2, and the parameter ε will be given a different definition from (9a). From the kinematic result (8) we now see that the governing equation is

$$F_t + \tfrac{1}{2}F_y^2 + 4F_{yyyy} - (1+\bar{l})F_{yy} + qF = 0, \tag{39}$$

a result identical to the curvature equation (23) when the scaling (9) is undone and R^{-1} replaced by q.

The linearized form of this equation was considered in §2, but we shall

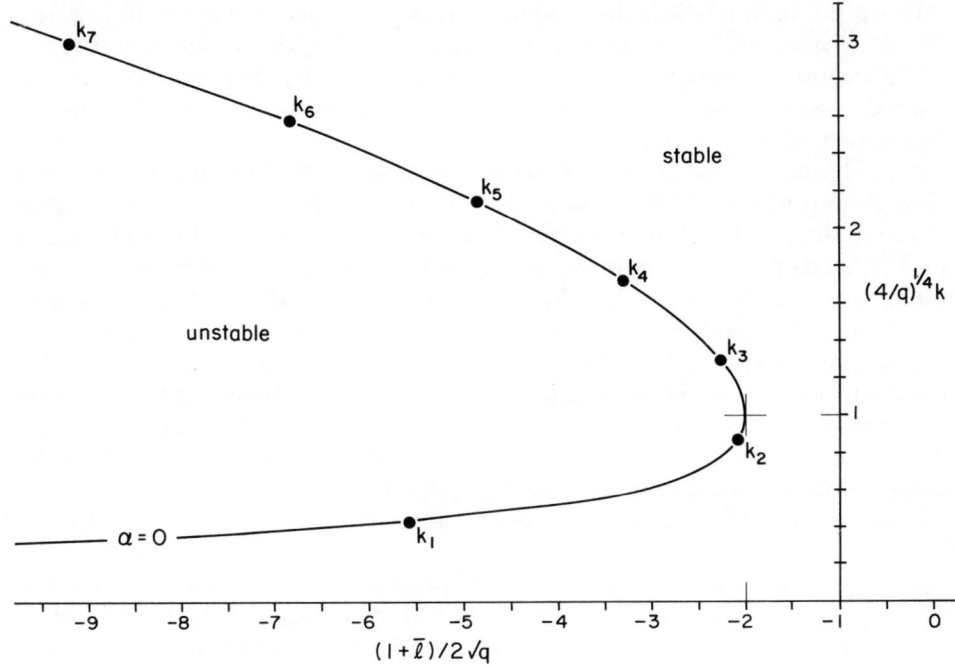

FIG. 6.8. *Linear stability regions for polyhedral flames, with admissible values of k.*

interpret the analysis differently. Rather than fixing $\bar{l}$ and determining the range of unstable wavenumbers for each q (there R^{-1}), we shall fix q and determine the range of unstable wavenumbers for each $\bar{l}$. Thus, with F proportional to $\exp(\alpha t + iky)$, the stability boundary $\alpha = 0$ is seen to be the curve

$$4k^4 + (1 + \bar{l})k^2 + q = 0 \qquad (40)$$

shown in Fig. 8.

Not all values of k are admissible, however, because an integer number of wavelengths must fit around the burner rim. The length L of the circumference provides the definition of the small parameter ε, namely

$$\varepsilon = \frac{4\pi^2}{L^2}, \qquad (41)$$

and then the restriction is

$$k = k_N \quad \text{with } k_N = N\sqrt{\varepsilon}, \quad N = 1, 2, 3, \ldots . \qquad (42)$$

The positions of the corresponding points on the neutral curve depend on the value of q; in drawing Fig. 8 we took $q = 117\varepsilon^2$. The corresponding values of $\bar{l}$ then lie in the order $\bar{l}_2, \bar{l}_3, \bar{l}_4, \bar{l}_5, \bar{l}_1, \bar{l}_6, \bar{l}_7, \ldots$. For larger values of q there are more points on the lower branch of the curves: in general, for $q > 4M^4\varepsilon^2$ there are M.

Those points for which the (reduced) Lewis number of the mixture is less than $\bar{l}_N$ correspond to unstable modes. Indeed, each l_N determines a supercritical bifurcation corresponding to a stationary N-sided polyhedral flame. Such unimodal bifurcations are essentially the same as those considered earlier in the context of flame curvature.

Both $\bar{l}$ and q can be varied in an experiment by changing the mixture strength and flow rate. We have seen that changes in q will move the points corresponding to k_N along the neutral curve, altering the stability characteristics of the flame. Changes in $\bar{l}$ will do so too. For certain choices of q, two bifurcation branches merge, i.e. $l_M = l_N$ for some M, N. If $M = N+1$ a study of the solution for parameter values that are varied close to those required for merging describes the transition from an N-sided flame to an $(N+1)$-sided flame, or the reverse. This transition can be quasisteady involving intermediate structures that are neither N- nor $(N+1)$-sided, or that are unsteady, phenomena that are observed experimentally. If $M = 2N$ the solution on the merged branch corresponds to a spinning polyhedral flame.

Insofar as spinning flames are concerned, the most satisfactory case, from a mathematical point of view, is $M=2$, $N=1$, (i.e., $q = 16\varepsilon^2$), for then the merged branch is the rightmost one and presumably is accessible as the first manifestation of instability. However one- or two-sided polyhedra do not fit comfortably on a circle, so that is not a physically satisfying choice. The objection does not apply to the choice $M=6$, $N=3$ (i.e., $q = 1296\varepsilon^2$), but then the two branches originating at $\bar{l}_4$ and $\bar{l}_5$ lie to the right of the merged branch, and our elementary analysis can provide no evidence that the latter is accessible. Indeed, it must be unstable near the bifurcation point where the weakly nonlinear analysis is valid. Buckmaster does not resolve this difficulty but makes a favorable comparison of the solution with the physical flame. Clearly a stabilizing mechanism must be found before it can be convincingly argued that polyhedral flames are completely understood. The corrugations of physical flames have large amplitude and perhaps the associated curvature is stabilizing, in which case the phenomenon lies outside the scope of a weakly nonlinear analysis. The analysis near the bifurcation point will now be sketched.

To ensure that k_N and k_{2N} give the same $\bar{l}$, we must take

$$q = 16N^4\varepsilon^2, \qquad (43)$$

and then

$$\bar{l}_N = \bar{l}_{2N} = -1 - 20N^2\varepsilon. \qquad (44)$$

To determine the solution on the merged branch, we perturb q and $\bar{l}$ away from the values (43), (44) by $O(\delta^2)$, where δ is a small perturbation parameter. At the same time we write

$$F = \delta f, \qquad t = \frac{\tilde{t}}{\delta} \qquad (45)$$

and expand f in a power series in δ. The leading term is found to be of the form

$$f_0 = A(\tilde{t})e^{ik_N y} + \bar{A}(\tilde{t})e^{-ik_N y} + B(\tilde{t})e^{ik_{2N} y} + \bar{B}(\tilde{t})e^{-ik_{2N} y}, \quad (46)$$

where

$$\frac{\partial A}{\partial \tilde{t}} = -2k_N^2 \bar{A} B, \qquad \frac{\partial B}{\partial \tilde{t}} = \tfrac{1}{2} k_N^2 A^2 \quad (47)$$

if there is to be no secular term in the perturbation of f. Partial derivatives are used because A and B also depend on the slow time $\delta^2 t$; evolution on this scale determines the ultimate amplitude of the spinning flame, but we shall not pursue the matter here.

Equations (47) have solutions corresponding to

$$f_0 = A_0[\pm\sqrt{8}\sin(\omega\tilde{t} + \phi \pm k_N y) - \sin 2(\omega\tilde{t} + \phi \pm k_N y)] \quad \text{with } \omega = k_N^2 A_0 > 0; \quad (48)$$

here A_0 and ϕ are real and constant on the $\tilde{t}$-scale. These are waves traveling in the negative/positive y-direction with an amplitude-dependent phase speed $k_N A_0$. The shape of the wave resembles the sawtooth profile in Fig. 6, and Buckmaster has argued that the propagation speed is consistent with the rapid rotations seen in experiments.

5. Other cellular flames. So far we have been concerned with the evolution of the linear instabilities associated with values of $\bar{l}$ slightly less than -1. Various additional effects were incorporated into the basic nonlinear theory, and others could have been (Sivashinsky (1983)). Our final remarks are concerned with values of $\bar{l}$ slightly greater than -1, where the linear stability of the flame can be destroyed by hydrodynamic effects.

The weakly nonlinear description is now

$$F_\tau + \tfrac{1}{2}F_\eta^2 + 4F_{\eta\eta\eta\eta} - F_{\eta\eta} + \gamma \mathrm{P}\!\!\int_{-\infty}^{\infty} \frac{F_\eta(\bar{\eta}, \tau)}{\bar{\eta} - \eta} d\bar{\eta} = 0 \quad \text{with } \gamma = \frac{\sigma - 1}{2\pi\varepsilon^{3/2}}. \quad (49)$$

Comparison with (15) reveals that the sign of $F_{\eta\eta}$ has been changed, because now the definition

$$\varepsilon = 1 + \bar{l} \quad (50)$$

is needed to obtain a positive parameter; η and τ still have the definitions (9b, c). Without the integral term, $F \to 0$ as $\tau \to \infty$ whatever the initial conditions are, corresponding to linear stability; as before, the hydrodynamic effects (represented by the integral) are destabilizing. Michelson and Sivashinsky's computations show that a progressive wave, consisting of stationary cells, eventually forms provided the flame is not too large. For large flames, the chaotic cellular structure first found in §1 reasserts itself.

The shape of the progressive wave satisfies a much simpler equation in the limit $\gamma \to 0$, i.e. for significantly larger departures of $\bar{l}$ from -1 than of σ from 1. Evolution is then on the scales $\gamma\eta$, $\gamma^2\tau$ rather than η, τ, so that the fourth

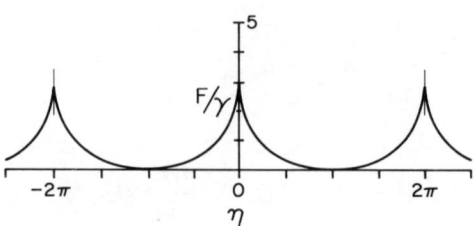

FIG. 6.9. *Stationary wrinkling of an otherwise stable plane flame due to hydrodynamic disturbances; possible outcome of Darrieus–Landau instability.*

derivative drops out. If a progressive wave is sought by setting $F_\tau = -V$ and if $F_{\eta\eta}$ is neglected (a valid step where the curvature is not large), the result is

$$\tfrac{1}{2}F_\eta^2 + \gamma \makebox[0pt]{$\displaystyle\int$}\rule[-0.5em]{0.6pt}{1.2em}_{-\infty}^{\infty} \left(\frac{1}{\bar\eta - \eta} - \frac{1}{\bar\eta}\right) F_\eta(\bar\eta)\, d\bar\eta = 0 \quad \text{with} \quad V = \gamma \makebox[0pt]{$\displaystyle\int$}\rule[-0.5em]{0.6pt}{1.2em}_{-\infty}^{\infty} \frac{F_{0\eta}(\bar\eta)}{\bar\eta}\, d\bar\eta, \qquad (51)$$

a nonlinear integral equation for the slope F_η.

The rough solution found by Sivashinsky was not very satisfactory and so McConnaughey, Ludford & Sivashinsky (1983) recently integrated the equation more accurately. A continuous periodic solution is shown in Fig. 9; the solution for any other period can be obtained from it by scaling η without changing V (there is no preferred wavelength in the linear theory). At the cusps, F_η has a logarithmic singularity, which makes the structure of the combustion field quite different from that of a Bunsen flame near its tip, for example. (Of course the singularity will be smoothed out by the neglected term $F_{\eta\eta}$, as Michelson and Sivashinsky's computations show.)

Of particular interest is the value

$$V = 1.4\gamma^2 \qquad (52)$$

obtained by McConnaughey, Ludford and Sivashinsky. For an expansion ratio of $\sigma = 5$, this leads to a flame speed that is 1.6 times the plane adiabatic value, a result in surprisingly good agreement with measured values (which range between 1.5 and 2). The theory, which assumes $\sigma - 1$ to be small, is not valid for such large expansion ratios, but nevertheless makes an accurate prediction.

LECTURE 7

Pulsating Flames

In § 5.3 it was found that plane NEFs of sufficiently large Lewis number are unstable. Since Im $(\alpha) \neq 0$ on the stability boundary, the instability is likely to result in either a pulsating flame or a flame that supports traveling waves. Such flames are the subject of this lecture.

One difficulty that immediately confronts us is that, in contrast to the ubiquitous nature of cellular instabilities, pulsating instabilities are not ordinarily seen. The reason seems to be the large values of $\mathscr{L}$ needed; according to the theory, $Y_f l/T_b^2$ must exceed $\frac{32}{3}$ (or $4(1+\sqrt{3})$ if the disturbances are one-dimensional). There is evidence that fuel-rich hydrogen-bromine mixtures might attain such values since Golovichev, Grishin, Agranat & Bertsun (1978) obtained oscillations in a numerical study, but there is no similar evidence for more commonplace gas mixtures.

For this reason we must turn from the commonplace and deal either with unusual combustible materials or else with special configurations, in order to uncover pulsating flames. Our discussion will start with thermites, which are solids that burn to form solids (a phenomenon that is appropriately called gasless combustion). There is no significant diffusion of mass, so that $\mathscr{L}$ is effectively infinite.

1. Solid combustion. Experiments by Merzhanov, Filonenko and Borovinskaya on thermites composed of niobium and boron revealed pulsating instabilities, as did numerical solutions obtained by Shkadinsky, Khaikin and Merzhanov. The latter examined the equations

$$\frac{\partial T}{\partial t} - \frac{\partial^2 T}{\partial x^2} = -\frac{\partial Y}{\partial t} = \Omega \tag{1}$$

where

$$\Omega = \mathscr{D} Y e^{-\theta/T} \quad \text{with} \quad \mathscr{D} = DM_r^{-2}, \tag{2}$$

and uncovered a critical value θ_c of the activation energy: for $\theta < \theta_c$ the propagation is steady, but for $\theta > \theta_c$ only pulsating propagation is seen. The prediction of pulsating propagation is consistent with the NEF analysis in § 5.3, where oscillatory instability was found for sufficiently large $\mathscr{L}$ in the limit $\theta \to \infty$; but activation-energy asymptotics has nothing to say about a phenomenon (here the switch to steady propagation) occurring at some finite value of θ. Even though it is not observed either experimentally or numerically for large enough θ, a steady wave can nevertheless be constructed by means of activation-energy asymptotics; we shall start our discussion by doing so.

The following boundary-value problem presents itself in a frame moving with the flame sheet:

$$\frac{dT}{dx} - \frac{d^2T}{dx^2} = -\frac{dY}{dx} = \mathcal{D}Ye^{-\theta/T}, \tag{3}$$

$$T \to T_f, \quad Y \to Y_f \quad \text{as } x \to -\infty, \qquad T \text{ bounded}, \quad Y \to 0 \quad \text{as } x \to +\infty. \tag{4}$$

The solution outside the flame sheet is

$$T = \begin{cases} T_f + Y_f e^x, \\ T_b = H_f, \end{cases} \quad Y = \begin{cases} Y_f \\ 0 \end{cases} \quad \text{for } x \lessgtr 0; \tag{5}$$

T is continuous across the flame sheet, as for the finite-$\mathcal{L}$ problem (cf. § 2.4), but Y jumps. As a consequence, the structure of the reaction zone is somewhat different.

The structure variable is, as usual,

$$\xi = \theta x \tag{6}$$

and

$$T_0 = T_b; \tag{7}$$

but we now find the equations

$$T_b^2 \frac{d^2\phi}{d\xi^2} = -\frac{dY_0}{d\xi} = \tilde{\mathcal{D}} Y_0 e^{-\phi} \quad \text{with } \tilde{\mathcal{D}} = \frac{\mathcal{D}e^{-\theta/T_b}}{\theta} \tag{8}$$

for the temperature perturbation $T_1 = -T_b^2 \theta$ and the leading term Y_0 in the mass-fraction expansion. Since both ϕ and Y_0 vanish as $\xi \to +\infty$, we have

$$Y_0 = -T_b^2 \frac{d\phi}{d\xi}, \tag{9}$$

so that only the temperature equation

$$\frac{d^2\phi}{d\xi^2} = -\tilde{\mathcal{D}} \frac{d\phi}{d\xi} e^{-\phi} \tag{10}$$

remains. To match with the solution outside the flame sheet, the usual boundary conditions

$$\phi = -\frac{Y_f \xi}{T_b^2} + o(1) \quad \text{as } \xi \to -\infty, \qquad \phi = o(1) \quad \text{as } \xi \to +\infty \tag{11}$$

must be applied. The first integral

$$\frac{d\phi}{d\xi} = \tilde{\mathcal{D}}(e^{-\phi} - 1) \tag{12}$$

of (10) is obtained by using the boundary condition (11b); then the boundary condition (11a) leads to the burning rate

$$M_r = \frac{\sqrt{D}\, T_b e^{-\theta/2T_b}}{\sqrt{Y_f \theta}}. \tag{13}$$

This result has apparently not been written down before, but others have derived it (e.g. Peters (1982)).

2. The delta-function model. We have already noted that there is little point in investigating the stability of this solution using activation-energy asymptotics. Instead, following Matkowsky and Sivashinsky, we shall use a delta-function model suggested by the asymptotics above. Thus, the strength of the delta function that replaces the Arrhenius term will be defined so that the mass flux through the flame sheet, in quite general circumstances, is

$$M = \frac{\sqrt{D T_*} e^{-\theta/2T_*}}{\sqrt{Y_f \theta}}; \qquad (14)$$

here T_* is the flame temperature. The dimensionless mass flux is then

$$\frac{M}{M_r} = \left(\frac{T_*}{T_b}\right) \exp\left[\frac{\theta(T_* - T_b)}{2 T_b T_*}\right], \qquad (15)$$

an expression that will be simplified before use. For θ large (but not necessarily infinite), the preexponential factor is not significant and can be replaced by 1; in addition, and consistently, for small deviations of T_* from T_b (such as occur in a linear stability analysis) the exponent can be replaced by $\theta(T_* - T_b)/2T_b^2$. The formula (15) then becomes

$$W = \frac{M}{M_r} = \exp\left[\frac{\theta(T_* - T_b)}{2 T_b^2}\right], \qquad (16)$$

since the dimensionless density may be taken to be 1; here W is the wave speed.

The Arrhenius term (2) is now replaced locally by

$$\Omega = Y_f W \Delta(n) \quad \text{with } n = x - F(0, 0, t), \qquad (17)$$

where Δ is the Dirac delta function and

$$x = F(y, z, t) \qquad (18)$$

is the flame sheet, written in a (fixed) coordinate system chosen so that the x-axis coincides with the normal at the point of interest at the instant considered. The equations

$$\frac{\partial T}{\partial t} - \nabla^2 T = -\frac{\partial Y}{\partial t} = \Omega, \qquad (19)$$

which generalize the one-dimensional ones (1), then show that the wave speed $-F_t(0, 0, t)$ is just W and that

$$\delta(T) = 0, \quad \delta\left(\frac{\partial T}{\partial n}\right) = -Y_f W, \quad \delta(Y) = Y_f. \qquad (20)$$

Precisely the same formulas can be obtained by applying activation-energy asymptotics to the flame sheet in the unsteady case (i.e. with $M \neq M_r$ and a flame temperature T_* within $O(\theta^{-1})$ of T_b) if it is assumed that there is no significant temperature gradient behind the flame sheet. What sets the delta-function method apart from activation-energy asymptotics is that, in solving (19), θ is treated as a finite parameter, and there is no requirement for the outer solution to match the inner in the limit $\theta \to \infty$.

3. Stability of thermite flames. Linear stability of the plane wave is investigated in the manner of § 5.3 for a plane NEF. Disturbances proportional to $\exp(\alpha t + iky)$ are sought, resulting in the dispersion relation

$$(2\alpha + \Theta)\sqrt{1 + 4(\alpha + k^2)} = \Theta(2\alpha + 1) \quad \text{with} \quad \Theta = \frac{Y_f \theta}{2T_b^2}. \tag{21}$$

The neutral stability curve

$$\Theta = \frac{6k^2 + 2 + (2k^2 + 1)\sqrt{16k^2 + 5}}{4k^2 + 1} \tag{22}$$

is shown in Fig. 1; on it

$$\alpha = \pm \frac{i\Theta^{1/2}(1 + 4k^2)^{1/2}}{2} \tag{23}$$

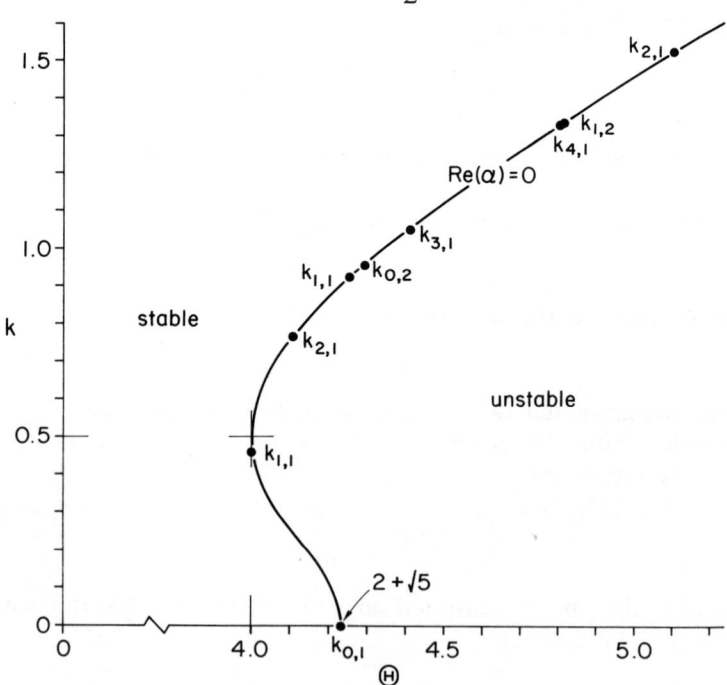

FIG. 7.1. Linear stability regions for thermite flames, with admissible values of k when confined to insulated circular cylinder. Labels to the right (left) of the neutral stability curve correspond to $a = 4$ (2).

is everywhere nonzero, i.e. the neutral modes are oscillatory. Resemblance to the right stability boundary for NEFs in Fig. 5.3 is striking. Note that the extreme value

$$\Theta_e = 4 \tag{24}$$

is large enough to give credence to the notion that most of the chemical reaction is confined to a thin sheet.

The results suggest that, for the one-dimensional equations (1), steady propagation is possibly only if

$$\Theta < \Theta_c = 2 + \sqrt{5}; \tag{25}$$

otherwise pulsating combustion occurs. This is in agreement with the experimental and numerical results cited at the beginning of § 1. Additional evidence is afforded by Matkowsky and Sivashinsky's demonstration that the passage of Θ through Θ_c gives a supercritical Hopf bifurcation.

Traveling-wave instabilities, rather than pulsations, will occur if disturbances of nonzero wavenumber are permitted by the lateral boundary conditions. The effect is strikingly seen for propagation through an insulated circular cylinder of thermite (Sivashinsky (1981)). Now disturbances proportional to $e^{in\phi}J_n(kr)$ are sought, where r, ϕ are polar coordinates in the y, z-plane and n is a nonnegative integer. Once more the dispersion relation (21), the neutral stability curve (22) and the neutral-mode frequency (23) are obtained, but now not all values of k are admissible. Thermal insulation of the surface $r = a$ requires

$$J'_n(ka) = 0, \quad \text{i.e.} \quad ka = j'_{n,m} \tag{26}$$

in a standard notation for zeros of derivatives of Bessel functions. The first seven allowable values of ka are

$$j'_{0,1} = 0, \quad j'_{1,1} = 1.84, \quad j'_{2,1} = 3.05, \quad j'_{0,2} = 3.83,$$

$$j'_{3,1} = 4.20, \quad j'_{4,1} = 5.32, \quad j'_{1,2} = 5.33 \tag{27}$$

according to Olver (1964). Some of the eigenvalues

$$k_{n,m} = \frac{j'_{n,m}}{a} \tag{28}$$

are marked on the neutral stability curve in Fig. 1 for $a = 2$ and 4. These two cases illustrate the general movement of the $k_{n,m}$-points down the curve as a increases.

Consider now what happens for cylinders of different size as Θ is increased up to the first onset of instability. (This is not easily done in a practical context.) Only the discrete points on the neutral stability curve corresponding to the values (28) of k are relevant, and which mode will be triggered first depends on the value of a. As Θ is further increased the mode becomes an admissible instability that develops a definite nonlinear form of the same general character.

For
$$a < 2.05 \qquad (29)$$
the point $k_{1,1}$ lies to the right of $k_{0,1}$ on the neutral curve; this is exemplified by $a = 2$ in Fig. 1. The first manifestation of instability will be plane pulsations of frequency
$$\omega = \pm \frac{\sqrt{\Theta_c}}{2}; \qquad (30)$$
the corresponding expressions for flame temperature and location are
$$T_* = T_b + \varepsilon \cos \omega t, \qquad x_* = -t - \left(\frac{8\varepsilon\omega}{Y_f}\right) \sin \omega t, \qquad (31)$$
where ε is the (linear) disturbance amplitude. The speed of the flame is greater or less than 1 accordingly as its temperature is greater or less than T_b. The temperature gradient
$$\varepsilon \operatorname{Re}\left[(\cos \omega t - i \sin \omega t)(1 - \sqrt{1 + 4i\omega})\right]/2 \qquad (32)$$
behind the flame does not vanish but fluctuates about zero (cf. the remark at the end of § 2).

The model is based on the assumption that the reactant is consumed completely, but this is not the case in practice. Indeed, it is sometimes possible to propagate a flame through the same material twice. It is to be expected that fluctuations in temperature gradient at the reaction front will result in a layered burnt state; in the context of activation-energy asymptotics, negative temperature gradients behind the reaction zone permit reactant leakage. Merzhanov, Filonenko and Borovinskaya noted layered structure in burnt thermites, with a layer for each pulsation.

A quite different phenomenon occurs if the first manifestation of instability is associated with $j'_{1,1}$, as is the case for
$$2.05 < a < 4.88; \qquad (33)$$
now $k_{1,1}$ is the leftmost point on the neutral curve, as is exemplified by $a = 4$ in Fig. 1. The expressions (31) are replaced by
$$T_* = T_b + \varepsilon \cos(\omega t + \phi)J_1(k_{1,1}r), \qquad x_* = -t - \left(\frac{8\varepsilon\omega}{Y_f}\right) \sin(\omega t + \phi)J_1(k_{1,1}r), \qquad (34)$$
where now
$$\omega = \pm \tfrac{1}{2}\Theta_{1,1}^{1/2}(1 + 4k_{1,1}^2)^{1/2}. \qquad (35)$$
The isotherms of the flame temperature are shown in Fig. 2; these spin, in either direction, with the frequency (35), producing a single hot spot traveling in a helical path on the surface of the cylinder. Such hot spots were observed by Merzhanov, Filonenko and Borovinskaya.

As the radius a increases beyond 4.88 for a certain interval, $k_{2,1}$ becomes the leftmost point on the neutral curve; there are then two spinning hot spots

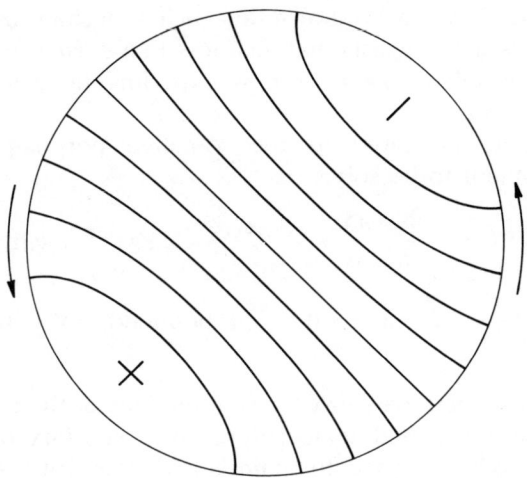

FIG. 7.2. *Isotherms of spinning thermite flame, with* $+/-$ *denoting the hot/cold sides.*

at opposite ends of a diameter. Hot spots can also occur in the interior, for example for $k_{1,2}$, which is the leftmost point for an interval of still larger values of a.

Other cross sections give rise to their own distinctive sets of admissible wavenumbers and isotherm patterns. For rectangles, judicious choice of proportion leads to 2 or even 3 modes simultaneously characterizing the onset of instability. A small change in the proportion will cause the corresponding eigenvalue to split, leading to secondary or tertiary bifurcations. Matkowsky & Olagunju (1982) have carried out the unimodal bifurcation analysis for circular cross sections (albeit in the different but related case of finite Lewis number); Margolis & Matkowsky (1983) have considered the multimodal analysis for rectangular cross sections.

The stability boundary identified here, coupled with the nature of the instabilities, shows that we are dealing with the analog of the right stability boundary in the NEF analysis. This suggests that SVFs, which lie between NEFs (with $\mathscr{L}$ close to 1) and thermite flames (with $\mathscr{L} = \infty$), should also exhibit pulsations as their instability; however, for $\mathscr{L} > 1$ their disturbances grow monotonically. Resolution of this apparent contradiction undoubtedly lies in the result (30) which suggests that the SVF analysis, by restricting attention to evolution on the time scale $t = O(\Theta)$, filters out a pulsating mode. In fact, Rogg (1982) has reported numerically determined pulsations for flames that would otherwise be candidates for SVF analysis.

4. Flames anchored to burners. We now turn our attention to gases, which necessarily have finite Lewis numbers, almost invariably lying to the left of the right stability boundary in Fig. 5.3. The problem is to find a mechanism that will shift this bounary to the left, making it accessible to mixtures of practical interest. Joulin and Clavin have shown that such a mechanism is the distributed heat loss of § 3.2, which suggests that heat loss to a burner anchoring the flame will have the same effect. There is now experimental evidence that burner flames can indeed pulsate.

Consider a flame anchored to the so-called porous-plug burner. The mathematical problem to be solved is

$$\frac{\partial T}{\partial t} + \frac{\partial T}{\partial x} - \nabla^2 T = -\frac{\partial Y}{\partial t} - \frac{\partial Y}{\partial x} + \mathscr{L}^{-1}\nabla^2 Y = \mathscr{D} Y e^{-\theta/T} \quad \text{with } \mathscr{D} = DM_r^{-2}, \quad (36)$$

$$T = T_s, \quad Y - \mathscr{L}^{-1}\frac{\partial Y}{\partial x} = J_s \quad \text{at } x = 0, \quad T \text{ bounded}, \quad Y \to 0 \quad \text{as } x \to +\infty. \tag{37}$$

Here M_r is the prescribed mass flux through the face of the plug at $x = 0$, while T_s and J_s are the prescribed temperature and mass-flux fraction there. (In practice, cooling coils are used to maintain T_s constant.) The physical idea underlying the boundary condition (37b) is that the porous surface inhibits the flux of reaction products, so that the flux fractions of the mixture as supplied by the burner are identical to the mass fractions in the reservoir supplying the burner.

Analysis of the steady problem in the limit $\theta \to \infty$ proceeds as in § 2.4 for the unbounded flame, except that the burnt-gas temperature T_b is the fundamental quantity to be determined, not M_r. Integration of the steady version of (36a) from $x = 0$ to ∞ yields

$$T'_s = T_s + J_s - T_b, \tag{38}$$

i.e. the heat received by the plug in terms of T_b. This enables us to write the solution in the form

$$T = \begin{cases} T_b - J_s + (T_s + J_s - T_b)e^x, \\ T_b, \end{cases} \quad Y = \begin{cases} J_s - J_s^{1-\mathscr{L}}(T_s + J_s - T_b)^{\mathscr{L}} e^{\mathscr{L}x} \\ 0 \end{cases} \quad \text{for } x \lessgtr x_*, \tag{39}$$

where

$$x_* = \ln\left[\frac{J_s}{(T_s + J_s - T_b)}\right]. \tag{40}$$

Consistency requires x_* to lie between 0 and ∞, so that we must have

$$T_s < T_b < T_s + J_s. \tag{41}$$

The right inequality shows that T'_s is necessarily positive, i.e. the plug must be a heat sink. Finally, a flame-sheet analysis gives

$$M_r = \frac{\sqrt{2\mathscr{L}D}\, T_b^2 e^{-\theta/2T_b}}{J_s \theta}, \tag{42}$$

from which T_b can be determined. (The result (2.43) is recaptured on replacing J_s with Y_f.) Ferguson and Keck have made satisfactory comparisons between experiment and a theoretical result essentially equivalent to the determination

(40) of the stand-off distance x_* as a function of the injection rate M_r. (Note that M_r is also used in making x_* nondimensional.)

The inequalities (41) define limits on M_r. When the injection rate is decreased (increased) beyond its limiting value a surface (remote) flame is obtained, requiring a different asymptotic analysis. We shall be concerned only with injection rates within the limits.

5. Stability of burner flames. In considering the stability of the solution in the last section it is natural to turn to a NEF analysis. However, such an analysis requires not only that $\mathscr{L}$ be sufficient close to 1, but also that the boundary conditions permit $T+Y$ to be constant to leading order. In general, the conditions (37) do not satisfy the second requirement.

One way out of the dilemma is to abandon activation-energy asymptotics and adopt a suitable modification of the delta-function model used in the discussion of thermites. (The strength of the delta function is again proportional to $\exp(-\theta/2T_*)$.) Such an approach was used by Margolis, who, by means of a numerical investigation of the complicated dispersion relation obtained from a linear stability analysis, was the first to demonstrate the leftward displacement of the stability boundary alluded to at the beginning of the last section. He also carried out a complete numerical simulation of a fuel-rich hydrogen-oxygen flame, thereby demonstrating pulsations; these were apparently confirmed by experiments performed at Sandia–Livermore, although there is no published account of them.

The NEF requirement that $T+Y$ should be constant to leading order is a sufficient but not necessary condition for the flame temperature to vary by $O(\theta^{-1})$ only. Thus, Buckmaster (1983b) also approached the question by means of a delta-function model, but then, a posteriori, justified the model through activation-energy asymptotics. This amounts to identifying the circumstances under which the dispersion relation is asymptotically meaningful, and for which there is a flame structure linking the states on the two sides of the flame sheet obtained by the equivalent jump conditions. Some care is necessary because of the sensitivity of the solution to variations in flame temperature; the disturbance field is $O(1)$ on the θ-scale, which entails calculating the flame-temperature perturbations correct to $O(\theta^{-1})$. In turn, this entails deriving jump conditions (from the flame-sheet structure) correct to the same order. Buckmaster found

$$\delta(T) = -\delta(Y) = \frac{\varepsilon T_{*1}}{\theta}, \tag{43}$$

$$\delta\left(\frac{\partial T}{\partial x}\right) = -\delta\left(\frac{\partial Y}{\partial x}\right) + \frac{lJ_s W}{\theta} = -J_s W + \frac{q}{\theta} \quad \text{with } W = \frac{1+\varepsilon T_{*1}}{2T_b^2}. \tag{44}$$

Here l has the same meaning (3.34) as in NEF analysis, ε is the small parameter characterizing the size of the disturbance, the flame temperature is $T_b + \varepsilon T_{*1}/\theta$, and q is a quantity (calculated from details of the flame structure) that is never needed. The term $-J_s W$ is the linearized form of the term $-Y_f W$ appearing in the jump condition (20b), with J_s replacing Y_f.

With the jump conditions in hand, it is a straightforward matter to carry out the analysis of unstable disturbances proportional to $\exp(\alpha t + iky)$. Circumstances justifying the delta-function model are then found to be

$$\theta e^{-\kappa x_*} = O(1) \tag{45}$$

where κ has the definition (5.33). An asymptotically self-consistent dispersion relation

$$2\kappa^2(1-\kappa^2) + \bar{l}(1+\kappa)[(1-\kappa)^2 - 4k^2] = 8\Theta\kappa^3 e^{-\kappa x_*} \tag{46}$$

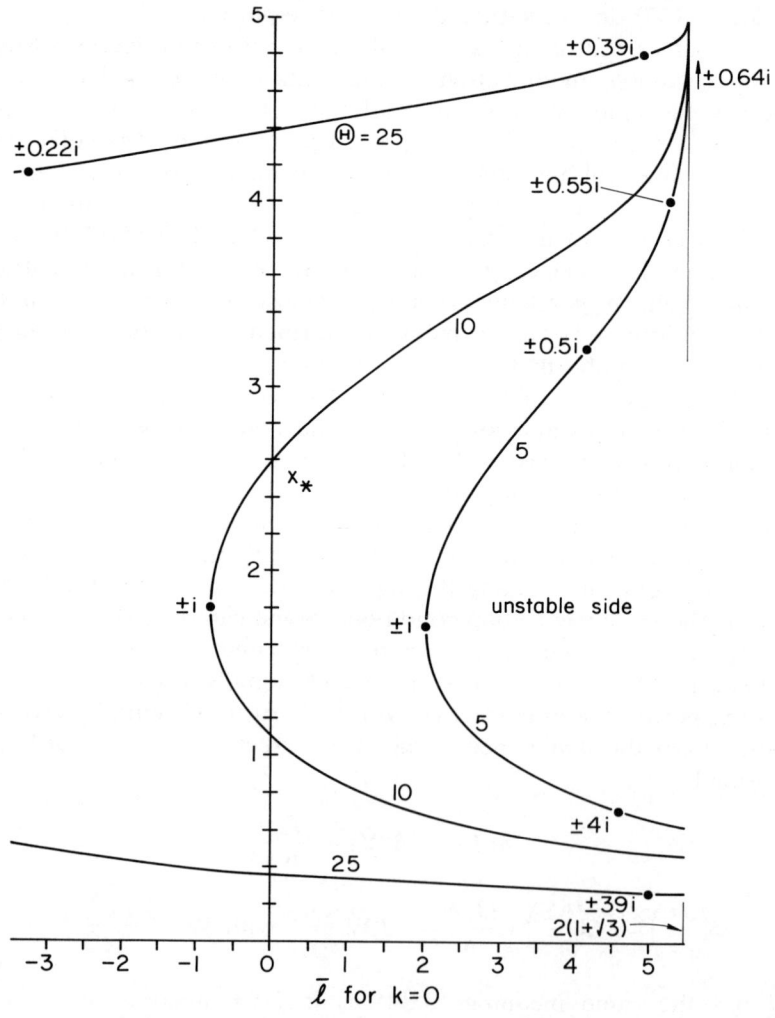

FIG. 7.3. *Displacement of the right stability boundary in Fig. 6.8 due to the presence of a burner, for various values of θ. The purely imaginary numbers are values of α.*

is obtained, the corresponding combustion field matching the structure used to deduce the jump conditions (43), (44); here $\bar{l}$ and Θ have the definitions (4.47) and (21b) with J_s in place of Y_f.

A similar dispersion relation (but free of θ and with $k=0$) was derived by Matkowsky and Olagunju for a somewhat artificial burner whose boundary conditions are compatible with NEF analysis. They do not discuss the full ramifications of their results.

What emerges from Buckmaster's analysis is essentially a NEF. To get a hint of this, note that the boundary conditions imply that the disturbance satisfies

$$T_1 + Y_1 = \mathscr{L}^{-1}\frac{\partial Y_1}{\partial x} \quad \text{at } x = 0. \tag{47}$$

Now the disturbance field decays rapidly ahead of the flame sheet, because κx_* is logarithmically large in θ. It follows that $\partial Y_1/\partial x$ is small at the plug, small enough for the condition (47) and the near-equality of thermal and mass diffusions to ensure that $T_1 + Y_1$ is, at most, $O(\theta^{-1})$ throughout the combustion field.

Finally, it should not be thought that the requirement (45) is a constraint on x_*. Insofar as the right stability boundary is concerned, the relation (46) implies that as x_* is decreased, Re (κ) increases (through an increase in the frequency of pulsations) so as to keep the term on the right balanced. When x_* is $O(1)$ the frequency is logarithmically large in θ.

The displacement of the right stability boundary is illustrated by Fig. 3, which shows how the point $k=0$ on it varies with x_*. (Note that $\bar{l} \to 2(1+\sqrt{3})$ as $x_* \to \infty$, in agreement with the result in § 5.3 for unbounded flames.) As x_* is decreased the boundary first moves to the left, by an amount that increases with Θ. But eventually this motion is halted and the boundary moves back to the right.

6. Pulsations for rear stagnation-point flow.

It seems likely that there are other mechanisms that will make the pulsations accessible. One that has been suggested by theory (but not confirmed experimentally) is negative strain, such as is experienced by a flame in a rear stagnation-point flow.

For moderate Reynolds numbers, flames can be stabilized behind the closed laminar wake at the rear of a thin plate or rod. Figure 4 shows the configuration, which Mikolaitis & Buckmaster (1981) treated with a NEF formulation based on the equations

$$\left(\frac{\partial}{\partial t} + \varepsilon y\frac{\partial}{\partial y} - \frac{\partial^2}{\partial y^2}\right)(T, h) = l\frac{\partial^2(0, T)}{\partial y^2} \tag{48}$$

and the flame sheet conditions (4.27)–(4.29). These equations are the unsteady version of (4.40) with the sign of ε changed because we are dealing with a rear instead of a front stagnation point. The problem is completed by the boundary

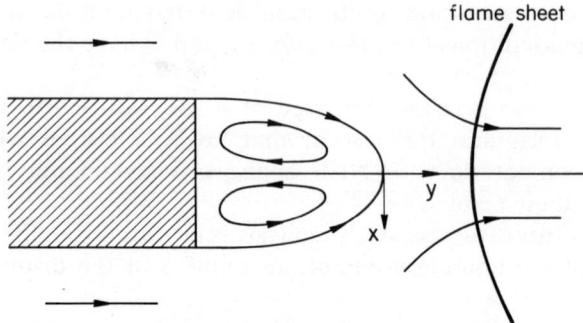

FIG. 7.4. NEF in a rear stagnation-point flow.

conditions

$$T = T_f, \quad h = 0 \quad \text{at } y = 0, \tag{49}$$

the latter corresponding to the prescription $Y = Y_f$ at $y = 0$.

The steady solution can be written in a closed form similar to that in § 4.5 for a front stagnation point. Its stability to one-dimensional disturbances was carefully explored using a combination of Galerkin's method and the method of weighted residuals. The results are illustrated by the two curves in Fig. 5, which shows variations in the stand-off distance y_* of the flame with (negative)

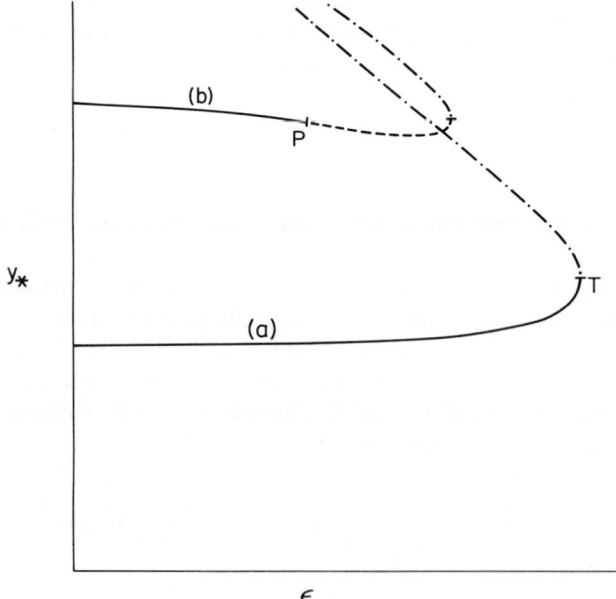

FIG. 7.5. Variation of stand-off distance y_* with straining rate ε for NEFs in rear stagnation-point flows.

straining rate. All responses have the form of a backward C, so that there is a maximum straining rate beyond which the flame must blow off.

For values of $\bar{l}\ (=lY_f/2T_b^2)$ less than about 0.91 the response is characterized by the curve (a); the lower branch is stable and the upper branch is unstable, with a real eigenvalue crossing through the origin as the turning point T is traversed. For other values of $\bar{l}$, as characterized by the curve (b), part of the lower branch is also unstable, with a complex conjugate pair of eigenvalues crossing the imaginary axis as the point P (for pulsations) is traversed. This raises the possibility that, for sufficiently large values of the Lewis number, blow-off will in practice be preceded by pulsations. (The critical value of $\bar{l}$ is quite accessible.)

LECTURE 8

Counterflow Diffusion Flames

The fundamental characteristic of diffusion flames is that the two reactants, fuel and oxidizer, are supplied in different parts of the combustion field, so that they must come together and mix by diffusion before reaction can take place. Counterflowing streams provide one method of bringing them together; the resulting diffusion flame, whose main properties were established by Liñán, is the subject of this lecture.

1. Basic equations. Consider the combustion field sketched in Fig. 1. A stream of gas containing the oxidant $Y_1 = X$ flows to the left and impinges on a stream containing the fuel $Y_2 = Y$ that flows to the right, forming a stagnation point at the origin. The flow field is

$$(u, v) = 2(-x, y) \tag{1}$$

under the constant-density approximation; here a constant of proportionality $\varepsilon/2$, where ε is the straining rate, has been absorbed into the length unit (which is used to define M_r). For such a flow, it is possible for the combustion field to be stratified in the y-direction, with a flat flame sheet at $x = x_*$. The temperature and mass fractions are then functions of x and t alone satisfying, for unit Lewis numbers,

$$\mathbb{L}(T) = -2\mathbb{L}(X) = -2\mathbb{L}(Y) = \Omega \quad \text{with } \Omega = \mathcal{D}XYe^{-\theta/T}; \tag{2}$$

here

$$\mathbb{L} \equiv \frac{\partial}{\partial t} - 2x\frac{\partial}{\partial x} - \frac{\partial^2}{\partial x^2} \tag{3}$$

and the boundary conditions are

$$T \to T_{f-}, \quad X \to 0, \quad Y \to Y_f \quad \text{as } x \to -\infty,$$
$$T \to T_{f+}, \quad X \to X_f, \quad Y \to 0 \quad \text{as } x \to +\infty. \tag{4}$$

The coefficient $\mathcal{D}$ is proportional to $1/\varepsilon$, so that an increase in the straining rate causes a decrease in $\mathcal{D}$. It is the response of the combustion field to variations in $\mathcal{D}$ that is of principal interest.

Since both Lewis numbers have been taken equal to 1, there are two Shvab–Zeldovich variables (§ 2.2), namely

$$T + 2X = T_\pm + Z_-, \qquad T + 2Y = T_\pm + Z_+, \tag{5}$$

where

$$T_\pm = \tfrac{1}{2}[T_{f+}\operatorname{erfc}(-x) + T_{f-}\operatorname{erfc}(x)], \quad Z_- = X_f\operatorname{erfc}(-x), \quad Z_+ = Y_f\operatorname{erfc}(x). \tag{6}$$

Both these variables are annihilated by $\mathbb{L}$ and satisfy the boundary conditions

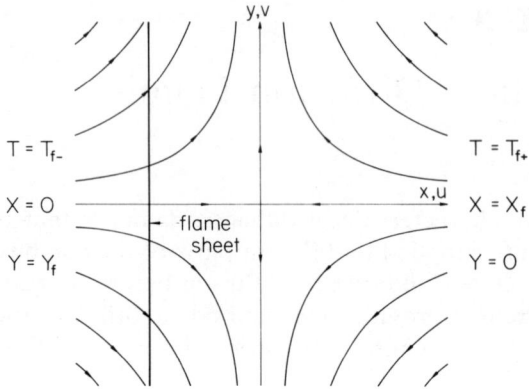

FIG. 8.1. *Notation for the counterflow diffusion flame.*

(4). We are left with a problem for the temperature alone, when X and Y are hereby suppressed in favor of T. Note that no assumption of steadiness has been made, but these relations can only be applied to unsteady problems for which the initial conditions satisfy the relations. We shall, however, be concerned solely with steady solutions until § 5.

When

$$\mathscr{D}e^{-\theta/T} \to 0 \qquad (7)$$

there is no chemical reaction, even if reactants are present, and the combustion is said to be frozen. There are two ways in which this can be brought about:
  (i) $\mathscr{D} \to 0$, the small Damköhler-number limit;
  (ii) $\mathscr{D} \sim e^{\theta/T_*}$ as $\theta \to \infty$ with $T < T_*$, a creature of activation-energy asymptotics.

In case (i) the entire combustion field is frozen, whereas in case (ii) only that portion where T falls below T_* is frozen.

When

$$\mathscr{D}e^{-\theta/T} \to \infty \qquad (8)$$

there is nothing to balance an infinite reaction rate outside vanishingly thin layers (spatial or temporal), and so

$$XY \to 0. \qquad (9)$$

At least one of the reactants is not present, i.e. there is equilibrium. Here, as for frozen combustion, there is no reaction, but for quite a different reason.

There are two ways in which equilibrium can be achieved:
  (i) $\mathscr{D} \to \infty$, the large Damköhler-number limit;
  (ii) $\mathscr{D} \sim e^{\theta/T_*}$ as $\theta \to \infty$ with $T > T_*$, another creature of activation-energy asymptotics.

In case (i) the entire combustion field is in equilibrium, but in case (ii) only that portion where T rises above T_* is in equilibrium. Because of the nature of this limit, there is always the possibility of a reaction zone, known as a Burke–Schumann flame sheet, existing in the middle of an equilibrium region; that cannot happen in a frozen region.

Simple analytical treatments are possible in these two limits, frozen and equilibrium. So elementary is the first that we shall forego discussion of it, concentrating instead on the second, which is called the Burke–Schumann (equilibrium) limit after an early, basic combustion problem they considered (see § 10.2). Large values of $\mathcal{D}$ are easily obtained in practice; this is the reason the limit was originally introduced by Burke and Schumann.

To the left of the flame sheet at $x = x_*$ there is no oxidant, and to the right no fuel. The Shvab–Zeldovich relations (5) therefore give

$$T = T_\pm + \begin{cases} Z_-, \\ Z_+, \end{cases} \quad X = \begin{cases} 0, \\ \frac{1}{2}(Z_- - Z_+), \end{cases} \quad Y = \begin{cases} \frac{1}{2}(Z_+ - Z_-) \\ 0 \end{cases} \quad \text{for } x \lessgtr x_*, \quad (10)$$

and continuity across the flame sheet then requires $Z_- = Z_+$ there, i.e.

$$\operatorname{erf}(x_*) = \frac{Y_f - X_f}{Y_f + X_f} \quad (11)$$

which determines x_*. The flame sheet lies in the leaner stream: $x_* \lessgtr 0$ accordingly as $X_f \gtreqless Y_f$. The flame temperature is

$$T_* = \frac{T_{f-} X_f + T_{f+} Y_f + 2 X_f Y_f}{X_f + Y_f}; \quad (12)$$

the associated structure (35), (36) will be derived in § 3 (for $\theta \to \infty$).

2. The S-shaped burning response. The usual way of characterizing the solution is to plot variations in some significant parameter, such as the maximum temperature, with the Damköhler number. If $T_{f\pm}$, X_f, Y_f are such that the combustion generates a heat flux to both far fields, this response is S-shaped in the limit $\theta \to \infty$ (Fig. 2). Certain physical conclusions can then be drawn.

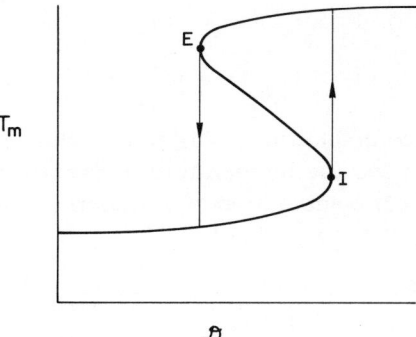

FIG. 8.2. *Ignition and extinction for S-shaped response.*

If the system is in a state corresponding to a point on the lower branch, and $\mathcal{D}$ is slowly increased, the solution can be expected to change smoothly until the point I is reached. Rapid transition to the upper branch will then presumably occur, corresponding to ignition. A subsequent slow decrease in $\mathcal{D}$ is likewise anticipated to produce a smooth decrease in burning rate until extinction occurs at E.

If one of the far fields loses heat, the response is monotonic, so that the phenomena of ignition and extinction are absent. For high activation energy the transition from a monotonic to an S-shaped response occurs when the temperature gradient on one side of the flame sheet is small, a case that is not difficult to analyze.

Assume, without loss of generality, that

$$T_{f+} > T_{f-}; \qquad (13)$$

then the small temperature gradient must be on the right of the flame sheet, i.e. on the hotter side. If it were on the colder side, the flame temperature (12) would be close to T_{f-}, which implies (in the limit of zero gradient)

$$X_f = \tfrac{1}{2}(T_{f-} - T_{f+}) < 0, \qquad (14)$$

an impossibility. To ensure that T_{f+} is close to the flame temperature, and hence that the temperature gradient is small on the hotter side, set

$$2Y_f = T_{f+} - T_{f-} + \frac{k}{\theta} \quad \text{with } k = \text{const.} \qquad (15)$$

In seeking an asymptotic solution as $\theta \to \infty$, we shall assume that equilibrium prevails for $x > x_*$ even though T does not rise above T_* by an $O(1)$ amount there, and check a posteriori that the solution thereby constructed is self-consistent.

In view of the assumption (15), the Shvab–Zeldovich relations (5b) becomes

$$T + 2Y = T_{f+} + \left(\frac{k}{2\theta}\right) \operatorname{erfc}(x) \qquad (16)$$

and, hence,

$$T = T_{f+} + \left(\frac{k}{2\theta}\right) \operatorname{erfc}(x) \quad \text{for } x > x_*, \qquad (17)$$

since $Y = 0$ there. To complete the description of the combustion field outside the reaction zone, we need the temperature in the frozen region ahead of the flame sheet, i.e. the linear combination of 1 and $\operatorname{erf}(x)$ that takes on the values T_{f-} at $x = -\infty$ and T_* at $x = x_*$; clearly

$$T = \frac{T_{f-}[\operatorname{erfc}(-x_*) - \operatorname{erfc}(-x)] + T_* \operatorname{erfc}(-x)}{\operatorname{erfc}(-x_*)} \quad \text{for } x < x_*. \qquad (18)$$

This, like the result (17), is correct to any order in θ^{-1}, provided T_* is determined to the same order. We shall only need leading-order accuracy in the result (18), so that taking

$$T_* = T_{f+} \tag{19}$$

is good enough. Determination of x_* (which need not be expanded) comes from analysis of the reaction zone, for which the leading-order result

$$X_* = \tfrac{1}{4}(T_{f-} - T_{f+}) \operatorname{erfc}(x_*) + \tfrac{1}{2}X_f \operatorname{erfc}(-x_*), \tag{20}$$

a consequence of the Shvab–Zeldovich relation (5a), is needed.

The appropriate variable in the reaction zone is

$$\xi = \theta(x - x_*), \tag{21}$$

so that coefficients in the layer expansion

$$T = T_{f+} - \theta^{-1} T_{f+}^2 \phi + \cdots \quad \text{with } \phi = \left(\frac{1}{T}\right)_1 \tag{22}$$

are considered to be functions of ξ. The Shvab–Zeldovich relation (5b) gives

$$Y = \tfrac{1}{2}\theta^{-1} T_{f+}^2 (\phi - \phi_*) + \cdots \quad \text{with } \phi_* = -\frac{k \operatorname{erfc}(x_*)}{2T_{f+}^2}, \tag{23}$$

so that the structure equation is

$$\frac{d^2\phi}{d\xi^2} = \tilde{\mathcal{D}}(\phi - \phi_*) e^{-\phi} \quad \text{with } \tilde{\mathcal{D}} = \frac{\mathcal{D} X_* e^{-\theta/T_{f+}}}{2\theta^2}. \tag{24}$$

Note that $\tilde{\mathcal{D}} = O(1)$ implies $\mathcal{D} e^{-\theta/T} = O(\theta^2)$ for $x > x_*$: there must be the equilibrium $Y = 0$ on the right of the flame sheet since otherwise the reaction term in the Y-equation could not be balanced there. In other words, our solution is self-consistent.

Equation (24) is precisely the structure equation for the premixed flames discussed in § 2.4. Since the gradient dT/dx vanishes on the right of the flame sheet, it determines the gradient on the left (as was explained in § 2.5). But the latter is known in terms of x_* from the expression (18), so the result is an equation for x_*, namely

$$\frac{2(T_{f+} - T_{f-})}{\sqrt{\pi}} \cdot \frac{e^{-x_*^2}}{\operatorname{erfc}(-x_*)} = \frac{T_{f+}^2 \sqrt{\mathcal{D} X_*} e^{-\phi_*/2}}{\theta e^{\theta/2T_{f+}}}. \tag{25}$$

When the definition (20) is used to eliminate X_*, equations (23) and (25) give ϕ_* (representing, when it is negative, the maximum temperature in the combustion field) and $\mathcal{D}$ as functions of x_*, i.e. the required relation between the maximum temperature and the Damköhler number.

The corresponding response curve is shown in Fig. 3 for several positive values of the constant k; when k is nonpositive, the maximum temperature is T_{f+} and not $T_{f+} - \theta^{-1} T_{f+}^2 \phi_*$. For k sufficiently small, the response is monotonic; otherwise it is S-shaped. Responses in the shape of an S appear to

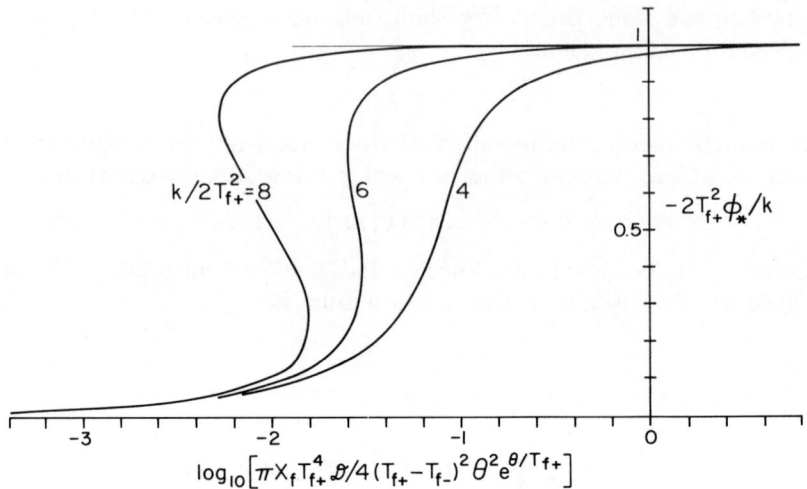

FIG. 8.3. *Steady-state responses when the temperature gradient is positive for $x<0$ and small for $x>0$, the latter being represented by k. Drawn for $X_f = \frac{1}{2}(T_{f+} - T_{f-})$.*

be associated with a flux of heat away from the flame sheet on both sides, but this has never been proved.

3. General extinction analysis. We turn now to the question of extinction in general, i.e. when there is an $O(1)$ heat flux away from the flame sheet in both directions. It is found that the flame temperature on the whole upper branch of the S-response differs by $O(\theta^{-1})$ from the Burke–Schumann value (12) at its end; our extinction analysis will use this fact. Because of the $O(1)$ drop in temperature away from the flame sheet, the combustion field on each side is now frozen; nevertheless, the leading-order solution outside the reaction zone is identical to the equilibrium solution constructed in the limit $\mathscr{D} \to \infty$ (§ 1). Thus, the results (10), which follow from equilibrium and the Shvab–Zeldovich relations, hold for T_0, X_0, Y_0. Finally, before considering the flame-sheet structure, we define x_* and T_* by the formulas (11) and (12).

In the reaction zone, the variable (21) and expansion (22) still apply, while the Shvab–Zeldovich relations gives

$$X = \tfrac{1}{2}\theta^{-1}T_*^2(\phi + A\xi) + \cdots, \qquad Y = \tfrac{1}{2}\theta^{-1}T_*^2(\phi + B\xi) + \cdots, \qquad (26)$$

where

$$A = \frac{(T_{f+} - T_{f-} + 2X_f)e^{-x_*^2}}{\sqrt{\pi}T_*^2} > 0, \qquad B = \frac{(T_{f+} - T_{f-} - 2Y_f)e^{-x_*^2}}{\sqrt{\pi}T_*^2} < 0; \qquad (27)$$

the signs of A and B follow from the requirement that the flame temperature be larger than the temperatures at both $-\infty$ and $+\infty$. The temperature equation therefore reduces to

$$\frac{d^2\phi}{d\xi^2} = \tilde{\mathscr{D}}(\phi + A\xi)(\phi + B\xi)e^{-\phi} \quad \text{with } \tilde{\mathscr{D}} = \frac{T_*^2 \mathscr{D} e^{-\theta/T_*}}{4\theta^3}, \qquad (28)$$

under the boundary conditions

$$\frac{d\phi}{d\xi} = -A + \cdots \quad \text{as } \xi \to -\infty, \qquad \frac{d\phi}{d\xi} = -B + \cdots \quad \text{as } \xi \to +\infty, \qquad (29)$$

which come from matching the leading terms in the expansions (26) with

$$X_0 = 0 \quad \text{for } x < x_*, \qquad Y_0 = 0 \quad \text{for } x > x_*. \qquad (30)$$

The problem (28), (29) determines the minimum value ϕ_m (i.e. the maximum temperature in the combustion field) as a function of $\tilde{\mathscr{D}}$. It can be reduced to canonical form by writing

$$\tilde{\phi} = \phi + \tilde{\gamma}\tilde{\xi}, \quad \tilde{\xi} = \frac{(A-B)\xi}{2} \quad \text{with } \tilde{\gamma} = \frac{A+B}{A-B}; \qquad (31)$$

we find

$$\frac{d^2\tilde{\phi}}{d\tilde{\xi}^2} = \tilde{\mathscr{D}}_e(\tilde{\phi}^2 - \tilde{\xi}^2)e^{-\tilde{\phi}+\tilde{\gamma}\tilde{\xi}} \quad \text{with } \tilde{\mathscr{D}}_e = \frac{4\tilde{\mathscr{D}}}{(A-B)^2} \qquad (32)$$

and

$$\frac{d\tilde{\phi}}{d\tilde{\xi}} = -1 + \cdots \quad \text{as } \tilde{\xi} \to -\infty, \qquad \frac{d\tilde{\phi}}{d\tilde{\xi}} = 1 + \cdots \quad \text{as } \tilde{\xi} \to +\infty. \qquad (33)$$

Various responses determined by numerical integration of the problem (32), (33) are shown in Fig. 4; these C-shaped curves correspond to the neighborhood of the point E in Fig. 2. Note that $A > 0, B < 0$ imply

$$-1 < \tilde{\gamma} < 1. \qquad (34)$$

Certain limiting forms of this problem as $\tilde{\mathscr{D}}_e \to \infty$ are of interest.

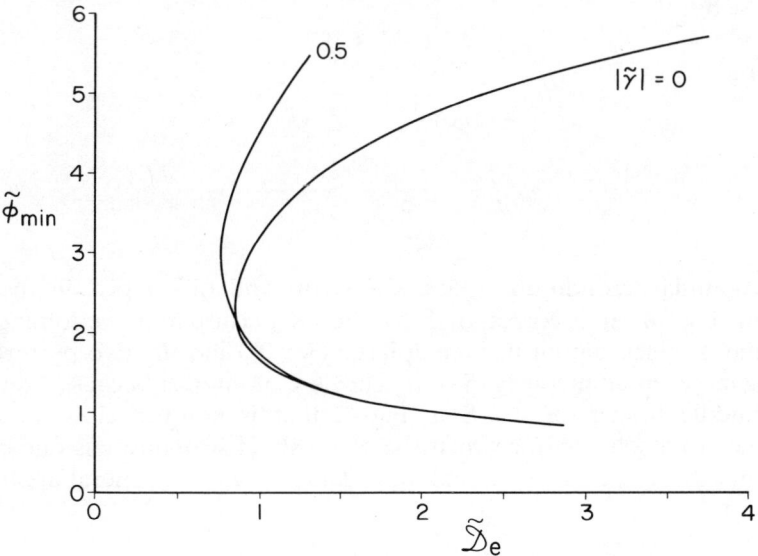

FIG. 8.4. Extinction curves. For $|\tilde{\gamma}| \geq 1$ no turning point is found numerically, a result that still lacks formal proof.

(i) The Burke–Schumann limit

$$\frac{d^2\hat{\phi}}{d\hat{\xi}^2} = (\hat{\phi}^2 - \hat{\xi}^2), \qquad \frac{d\hat{\phi}}{d\hat{\xi}} = \mp 1 + \cdots \quad \text{as } \hat{\xi} \to \mp\infty \tag{35}$$

is obtained by setting

$$\tilde{\phi} = \tilde{\mathcal{D}}_e^{-1/3}\hat{\phi}, \qquad \tilde{\xi} = \tilde{\mathcal{D}}_e^{-1/3}\hat{\xi}. \tag{36}$$

This is the structure problem for the equilibrium solution discussed at the end of § 1; it applies far to the right on the upper branch in Fig. 4. Existence and uniqueness have recently been proved by Holmes.

(ii) The so-called premixed-flame limit results from defining a small parameter ε by

$$\varepsilon^{1\mp\tilde{\gamma}} \ln\left(\frac{1}{\varepsilon}\right) = \mathcal{D}_e^{-1} \quad \text{according as } \tilde{\gamma} \gtrless 0 \tag{37}$$

and putting

$$\tilde{\phi} = \hat{\phi} - \ln\varepsilon, \qquad \tilde{\xi} = \hat{\xi} \mp \ln\varepsilon. \tag{38}$$

We find

$$\frac{d^2\hat{\phi}}{d\hat{\xi}^2} = 2(\hat{\phi} \mp \hat{\xi})e^{-\hat{\phi}+\tilde{\gamma}\hat{\xi}} \tag{39}$$

and the boundary conditions (35b); existence of the solution requires changing from one equation to the other when the sign of $\tilde{\gamma}$ is changed. The relevance of this structure problem to the lower branch in Fig. 4 is discussed below.

(iii) The so-called partial-burning limit corresponds to

$$\tilde{\gamma} = 0. \tag{40}$$

Defining ε by

$$\varepsilon^{-2} e^{-1/\varepsilon} = \mathcal{D}_e^{-1} \tag{41}$$

and setting

$$\tilde{\phi} = \hat{\phi} + \frac{1}{\varepsilon}, \qquad \tilde{\xi} = \hat{\xi} \tag{42}$$

give

$$\frac{d^2\hat{\phi}}{d\hat{\xi}^2} = e^{-\hat{\phi}} \tag{43}$$

and the boundary conditions (35a). Cases (ii) and (iii) apply on the lower branch in Fig. 4 and correspond to the X-perturbation becoming large ($\tilde{\gamma} > 0$), the Y-perturbation becoming large ($\tilde{\gamma} < 0$), and the two perturbations becoming large simultaneously ($\tilde{\gamma} = 0$). They are of interest because every point on the middle branch of the S in Fig. 2 that is not too close to E or I corresponds to a solution for which the flame-sheet structure has one of these three forms. As an example we shall now demonstrate the general applicability of case (iii).

4. Partial-burning branch. This part of the S is characterized by $O(1)$ values of X_* and Y_*, hence the term partial burning. It is convenient to prescribe the flame temperature T_* (also the maximum temperature) and calculate $\mathscr{D}$, rather than vice versa.

On either side of the flame sheet the leading-order temperature is

$$T_0 = T_{f\mp} + \frac{(T_* - T_{f\mp})\,\text{erfc}\,(\mp x)}{\text{erfc}\,(\mp x_*)} \quad \text{for } x \lessgtr x_*. \tag{44}$$

In order to determine x_*, we anticipate a conclusion to be drawn from the flame-sheet structure, to wit

$$\frac{dT}{dx}(x_* - 0) + \frac{dT}{dx}(x_* + 0) = 0; \tag{45}$$

this gives

$$\frac{\text{erfc}\,(x_*)}{\text{erfc}\,(-x_*)} = (T_* - T_{f+})(T_* - T_{f-}). \tag{46}$$

The Shvab–Zeldovich relations (5) then determine X_* and Y_*.

In the reaction zone, the variable (21) and expansion (22) still apply, so that the structure equation is

$$\frac{d^2\phi}{d\xi^2} = \tilde{\mathscr{D}} e^{-\phi} \quad \text{with } \tilde{\mathscr{D}} = \frac{\mathscr{D} X_* Y_* e^{-\theta/T_*}}{T_*^2 \theta}, \tag{47}$$

which is equivalent to (43). A single integration shows that the derivative $d\phi/dx$ takes equal but opposite values at $\xi = \pm\infty$, the origin of the relation (45). No information is obtained about $\mathscr{D}$, however, except that it is proportional to e^{θ/T_*} and is therefore a rapidly decreasing function of T_*; to determine it requires an examination of higher-order terms.

This structure is appropriate for the lower part of the middle branch. As T_* increases on moving up the branch, one mass fraction, X_* or Y_*, decreases to zero; therefore, a different structure takes over, characterized by $O(\theta^{-1})$ values of one of the mass fractions but still $O(1)$ values of the other. One or other of equations (39) then governs; again $\mathscr{D}$ is proportional to e^{θ/T_*}. Further increases in T_* towards the Burke–Schumann value (12) causes the remaining $O(1)$ mass fraction to decrease until the structure (32), (33) is attained. We forego further discussion of the middle branch since it corresponds to unstable solutions, our final topic.

5. Stability. The middle branch has long been believed to be unstable, but only recently has the matter been confirmed mathematically. To do so, the problem must be examined on a time scale that is relevant to the reaction zone, i.e. using a fast time

$$\tau = \theta^2 t. \tag{48}$$

The governing equations (2) show that

$$\frac{\partial T}{\partial \tau} = \frac{\partial X}{\partial \tau} = \frac{\partial Y}{\partial \tau} = 0 \tag{49}$$

everywhere on the two sides of the flame sheet: T, X, and Y are described to all orders by the steady state.

On the other hand, in the reaction zone the time derivatives are as important as the diffusion terms, so that (28a) becomes

$$\frac{\partial^2 \phi}{\partial \xi^2} - \frac{\partial \phi}{\partial \tau} = \tilde{\mathscr{D}}(\phi + A\xi)(\phi + B\xi)e^{-\phi}. \tag{50}$$

Correspondingly, (32a) is replaced by

$$\frac{\partial^2 \tilde{\phi}}{\partial \tilde{\xi}^2} - \frac{\partial \tilde{\phi}}{\partial \tilde{\tau}} = \tilde{\mathscr{D}}_e(\tilde{\phi}^2 - \tilde{\xi}^2)e^{-\tilde{\phi}+\tilde{\gamma}\tilde{\xi}} \quad \text{with } \tilde{\tau} = \frac{(A-B)^2 \tau}{4}. \tag{51}$$

Infinitesimal disturbances $e^{\lambda \tilde{\tau}} \hat{\phi}_1$ of the steady state $\tilde{\phi}_0$, which is the solution of the problem (32), (33), are then described by the eigenvalue problem

$$\hat{\phi}_1'' + [V(\tilde{\xi}) - \lambda]\hat{\phi}_1 = 0, \qquad \hat{\phi}_1 \to 0 \quad \text{as } \tilde{\xi} \to \pm\infty, \tag{52}$$

where

$$V = \tilde{\mathscr{D}}_e(\tilde{\phi}_0^2 - 2\tilde{\phi}_0 - \tilde{\xi}^2)e^{-\tilde{\phi}_0+\tilde{\gamma}\tilde{\xi}}. \tag{53}$$

(Attention is restricted to disturbances satisfying the Shvab–Zeldovich relations.)

This problem has been treated by Buckmaster, Nachman and Taliaferro (1983) who show, in particular, that the transition from stable solutions on the upper branch to unstable solutions on the middle branch occurs exactly at the static extinction point E in Fig. 2. (The rigorous part of the analysis is due to Taliaferro.) Stability on the upper branch is typified by the Burke–Schumann limit (35), for which the corresponding eigenvalue problem is

$$\hat{\phi}_1'' - (2\hat{\phi}_0 + \hat{\lambda})\hat{\phi}_1 = 0 \quad \text{with } \hat{\lambda} = \mathscr{D}_e^{-2/3}\lambda, \qquad \hat{\phi}_1 \to 0 \quad \text{as } \hat{\xi} \to \pm\infty. \tag{54}$$

Instability on the middle branch is exemplified by the premixed-flame and partial-burning limits (39) and (43). For the former the eigenvalue problem is

$$\hat{\phi}_1'' - [2(1 \pm \hat{\xi} - \hat{\phi}_0)e^{-\hat{\phi}_0+\tilde{\gamma}\tilde{\xi}} + \lambda]\hat{\phi}_1 = 0 \tag{55}$$

under the boundary conditions (54b); for the latter it is

$$\hat{\phi}_1'' + (e^{-\hat{\phi}_0} - \lambda)\hat{\phi}_1 = 0 \tag{56}$$

under the same boundary conditions.

Standard treatments show that the spectrum of λ is negative if $\hat{\phi}_0$ is everywhere positive, and that is ensured for the problem (54) by the property $\hat{\phi}_0 > |\hat{\xi}|$ of the steady state. Peters was the first to show, albeit numerically, that the problems (55) have eigenvalues with positive real part. (Peters writes equations in a slightly different way from ours; his m is equivalent to $2(1-|\tilde{\gamma}|)$ and so is positive.) Matalon and Ludford noted, in a different context, that the problem (56) has a positive eigenvalue. Indeed, the steady state is

$$\hat{\phi}_0 = \ln[2 \cosh^2 \tfrac{1}{2}(\hat{\xi} - \hat{\xi}_m)], \tag{57}$$

where $\hat{\xi}_m$ is an integration constant giving the location of the maximum temperature, so that

$$\lambda = \tfrac{1}{4}, \qquad \hat{\phi}_1 = \operatorname{sech} \tfrac{1}{2}(\hat{\xi} - \hat{\xi}_m) \tag{58}$$

are seen to be an eigenvalue and its corresponding eigenfunction.

6. The ignition point. The neighborhood of the point I in Fig. 2 can be analyzed by considering perturbations of the frozen solution that raise the maximum temperature by an $O(\theta^{-1})$ amount. A similar analysis in the context of spherical diffusion flames will be given in the next lecture, so we omit the discussion for counterflow flames. The stability of the lower branch has not yet been examined, but the conclusions must agree with the instability result for the partial-burning portion of the middle branch.

LECTURE 9

Spherical Diffusion Flames

Law has shown that the analysis of spherical diffusion flames is quite similar to that of counterflow diffusion flames, so that some explanation is needed for devoting a separate lecture to them. There are two good reasons. First, the constant-density approximation has been used throughout these lectures in discussing all but plane flames, so there is room for a problem which does not neglect variations in density. (Plane diffusion flames have to be chambered, i.e. the reactants must be supplied at finite locations, which leads to distracting complications.) Secondly, the spherical diffusion flame can lead to quite different (and unusual) responses. These arise in the technologically important application to the quasi-steady phase of fuel-drop burning, when a more realistic boundary condition than the conventional one is used.

1. Basic equations. The fuel is supposed to be supplied as a liquid at the surface of a sphere of radius a (Fig. 1), where the heat from a concentric flame sheet evaporates it. In turn, the flame sheet is sustained by the reaction of the gaseous fuel and oxidant in the ambient atmosphere. If the representative mass flux M_r is taken to be $\lambda/c_p a$, then a is the length unit. The density at infinity will be used for ρ_r.

We seek a spherically symmetric, steady solution of the full equations (2.18)–(2.20) under the assumption of unit Lewis numbers for the reactants. The equation of continuity integrates immediately to

$$r^2 \rho v = M \text{ (const.)}, \qquad (1)$$

where M is loosely called the burning rate. (The liquid fuel evaporates at the rate $4\pi M$, but the gaseous fuel only burns at that rate if there is no outflow of fuel at infinity.) Although the radial velocity v is not determined by this result, the thermal-chemical equations (2.20) are uncoupled from the fluid-mechanical equations (2.18b), (2.19) because, under steady conditions, they only involve the mass flux ρv, which is determined (as M/r^2). Once T is found, the density follows from

$$\rho T = T_\infty, \qquad (2)$$

and then v is known; the momentum equation (2.19) serves only to determine the pressure field.

We have to deal once more with equations (8.2), where now, however,

$$\mathbb{L} \equiv \frac{M}{r^2} \frac{d}{dr} - \frac{1}{r^2} \frac{d}{dr} r^2 \frac{d}{dr}. \qquad (3)$$

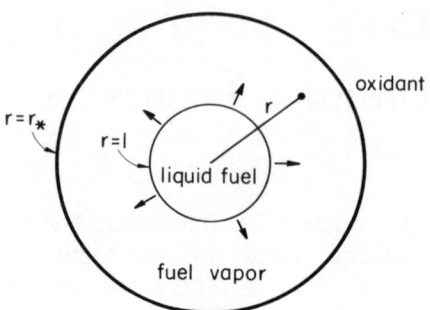

FIG. 9.1. *Burning fuel drop.*

The boundary conditions are

$$T = T_s, \quad \frac{dT}{dr} = ML, \quad X - M^{-1}\frac{dX}{dr} = 0, \quad Y - M^{-1}\frac{dY}{dr} = 1 \quad \text{at } r = 1, \tag{4}$$

$$T \to T_\infty, \quad X \to X_\infty, \quad Y \to 0 \quad \text{as } r \to \infty, \tag{5}$$

where T_s, L, T_∞, X_∞ are given. Prescription of the surface temperature T_s will later be replaced by the requirement of liquid-vapor equilibrium; the latent heat of evaporation L is positive; the conditions (4c, d) ensure that the sphere is a source of fuel, but neither a source nor sink of oxidant (or anything else); the prescribed oxidant fraction at infinity must of course satisfy $0 < X_\infty \le 1$; and the condition (5c) ensures that all the fuel has originated at the supply.

The sixth-order system of differential equations is therefore subject to seven boundary conditions, as a consequence of which M is determined as a function of $\mathscr{D}$ (depending also on T_s, L, T_∞, and X_∞). The maximum temperature could be used to characterize the solution, as for the counterflow diffusion flame, but we shall use M instead. The burning rate is of greater interest in applications, and it arises naturally in the analysis.

The Shvab–Zeldovich relations are

$$T + 2X = (T_s - L)(1 - e^{-M/r}) + (T_\infty + 2X_\infty)e^{-M/r} \equiv Z_-(M), \tag{6}$$

$$T + 2Y = T_\infty + (T_a - T_\infty)(1 - e^{-M/r}) \equiv Z_+(M), \tag{7}$$

where $T_a = T_s - L + 2$ is called the adiabatic flame temperature. For $T_\infty = T_a$, no heat is conducted to or from the ambient atmosphere when (as is usually the case) $Y = o(r^{-1})$ as $r \to \infty$; this conclusion follows from the relation (7), which gives

$$\lim_{r \to \infty}\left(4\pi r^2 \frac{dT}{dr}\right) = \lim_{r \to \infty}\left[4\pi r^2 (T_a - T_\infty)\left(\frac{-Me^{-M/r}}{r^2}\right)\right] = 4\pi M(T_\infty - T_a). \tag{8}$$

We further see that

$$T_a > T_\infty \tag{9}$$

will ensure the conduction of heat to the environment, the aim of combustion in practice.

The problem now reduces to one for T alone, namely

$$\mathbb{L}(T) = \mathcal{D}XYe^{-\theta/T}, \quad T = T_s, \quad \frac{dT}{dr} = ML \text{ at } r = 1, \quad T \to T_\infty \text{ as } r \to r_\infty, \quad (10)$$

where X and Y are to be suppressed in favor of T by means of the Shvab–Zeldovich relations (6), (7). We shall first derive the frozen ($\mathcal{D}e^{-\theta/T} \to 0$) and equilibrium ($\mathcal{D}e^{-\theta/T} \to \infty$) limits of the solution.

For frozen combustion, start by setting

$$\varepsilon^2 = \tfrac{1}{2}\mathcal{D}X_\infty e^{-\theta/T_\infty} \tag{11}$$

equal to zero. We find

$$T = T_s - L + Le^{M_w(1-1/r)} \quad \text{with } M_w \equiv \ln\left(1 + \frac{T_\infty - T_s}{L}\right) \tag{12}$$

where, since Y must be positive,

$$M_w > 0, \quad \text{i.e. } T_\infty > T_s. \tag{13}$$

This is the pure evaporation solution with no combustion; the environment must be hotter than the surface in order to supply the heat necessary to vaporize the liquid. (There is no condensation solution.) The result is not uniformly valid, since for $r = O(\varepsilon^{-1})$ the reaction term is comparable to the convection term in the equation (10a). The variable

$$R = \varepsilon r \tag{14}$$

leads to the expansion

$$T = T_\infty - \varepsilon M_w \frac{T_\infty - T_s + L - 2(1 - e^{-R})}{R} + \cdots. \tag{15}$$

The frozen limit is not an extinguished state, but rather one in which all the reaction takes place at essentially constant temperature far from the supply.

Turning now to the (Burke–Schumann) equilibrium limit, we use the Shvab–Zeldovich relations (6), (7) to obtain

$$\tag{16}$$

$$T = \begin{cases} Z_-(M_e), \\ Z_+(M_e), \end{cases} \quad X = \begin{cases} 0, \\ (1+X_\infty)e^{-M_e/r} - 1, \end{cases} \quad Y = \begin{cases} 1 - (1+X_\infty)e^{-M_e/r} \\ 0 \end{cases} \quad \text{for } r \lessgtr r_*,$$

where the boundary condition (10b) fixes

$$M_e \equiv \ln\left(1 + \frac{T_\infty - T_s + 2X_\infty}{L}\right) \tag{17}$$

and the continuity of X and Y across the flame sheet requires

$$r_* = \frac{M_e}{\ln(1+X_\infty)}. \tag{18}$$

The flame temperature, i.e. the common value to which T tends as $r \to r_* \pm 0$, is

$$T_* = T_\infty + \gamma X_\infty \quad \text{with} \quad \gamma = \frac{T_a - T_\infty}{1 + X_\infty}. \tag{19}$$

For consistency, the value (18) of r_* must be greater than 1, i.e.

$$T_\infty - T_s > (L-2)X_\infty. \tag{20}$$

While this condition is automatically met when the inequalities (9), (13b) are satisfied (as they are in the practically important case to which we shall limit our later discussion), it is of interest to note that there is a second limit, found by Buckmaster, when the condition is violated: The flame sheet lies at the surface itself, instead of some distance away, so that there is no oxidant-free region.

The flame-sheet structure in the Burke–Schumann limit is the same irrespective of whether the counterflow or spherical diffusion flame is being considered (cf. end of § 8.1). The continuity conditions

$$\delta(T) = \delta(X) = \delta(Y) = 0 \tag{21}$$

are required for a structure to exist, but the gradients of T, X and Y are different on the two sides of the flame sheet because it is a source of heat and a sink of both reactants. The Shvab–Zeldovich relations show, however, that

$$\delta\left(\frac{dT}{dr} + 2\frac{dX}{dr}\right) = \delta\left(\frac{dT}{dr} + 2\frac{dY}{dr}\right) = 0. \tag{22}$$

2. Nearly adiabatic burning. We turn now to the full response curve $M(\mathscr{D})$ in the limit $\theta \to \infty$, noting once again that the burning rate is a more convenient and significant parameter than the maximum temperature to characterize the solution. Here the response is S-shaped if T_s, L, T_∞ satisfy the inequalities (9) and (13b), i.e. when the heat flux (8) is to the environment, as required in practice, and the surface temperature is less than that of the ambient atmosphere. The transition from an S-shaped response to a monotonic one occurs, for large activation energy, when the temperature gradient (i.e. the heat flux) beyond the flame is small, so that it may be described by setting

$$T_a - T_\infty = \frac{k}{\theta} \quad \text{with} \quad k = \text{const.}; \tag{23}$$

here T_∞ and L are supposed fixed as T_s varies. The inequality (13b), which ensures a weak-burning branch for the response, requires

$$L < 2. \tag{24}$$

Note that the requirement (20) for the Burke–Schumann limit is automatically satisfied, ensuring a strong-burning branch of that form.

Equilibrium will be assumed behind the flame sheet even though the temperature does not rise significantly above T_* ($= T_*$, to leading order) there. The combustion is frozen between the surface and the flame sheet, because of the requirement (13b). The Shvab–Zeldovich relation (7) therefore gives us

$$T = T_\infty + \frac{k}{\theta}(1 - e^{-M/r}) \quad \text{for } r > r_*, \tag{25}$$

in view of the assumption (23), and the boundary conditions (10b, c) lead to

$$T = T_s - L + L e^{M(1 - 1/r)} \quad \text{for } r < r_*. \tag{26}$$

These formulas are correct to any order in θ^{-1}, provided that M is determined to the same order; we shall need only a leading-order determination of M, and that will be understood in what follows. Leading-order continuity of T now shows that

$$r_* = \frac{M}{M + \ln(L/2)}, \tag{27}$$

a result that exhibits the need for the inequality (24). To analyze the reaction zone we shall also need the leading-order result

$$X_* = (1 + X_\infty)e^{-M/r_*} - 1, \tag{28}$$

coming from the Shvab–Zeldovich relation (6).

Determination of M comes from analysis of the reaction zone, for which the appropriate variable is

$$\xi = \theta(r - r_*). \tag{29}$$

Coefficients in the layer expansion

$$T = T_\infty - \theta^{-1} T_\infty^2 \phi + \cdots \quad \text{with } \phi = \left(\frac{1}{T}\right)_1 \tag{30}$$

are considered to be functions of ξ. The Shvab–Zeldovich relation gives

$$Y = \tfrac{1}{2}\theta^{-1} T_\infty^2 (\phi - \phi_*) + \cdots \quad \text{with } \phi_* = -\frac{k(1 - e^{-M/r_*})}{T_\infty^2}, \tag{31}$$

so that the structure is governed by (8.24) with T_{f+} replaced by T_∞. As there, the structure equation provides an expression for the temperature gradient in the field (26) in front of the flame sheet, whence we find the equation

$$\frac{2M}{r_*^2} = \frac{T_\infty^2 \sqrt{\mathscr{D} X_*} e^{-\phi_*/2}}{\theta e^{\theta/2T_\infty}} \tag{32}$$

for M as a function of $\mathscr{D}$. This function can be defined parametrically, as for the counterflow diffusion flame at the end of § 8.2. Thus, M is given as a function of r_* by (27), and then $\mathscr{D}$ is determined as a function of r_* by (32) with the substitutions (28) and (31b) for X_* and ϕ_*.

The corresponding response curve is shown in Fig. 2 for several values of k. The curve joins the frozen limit (12b) to the Burke–Schumann limit (17), monotonically if k is not too positive but otherwise via an S. Responses in the

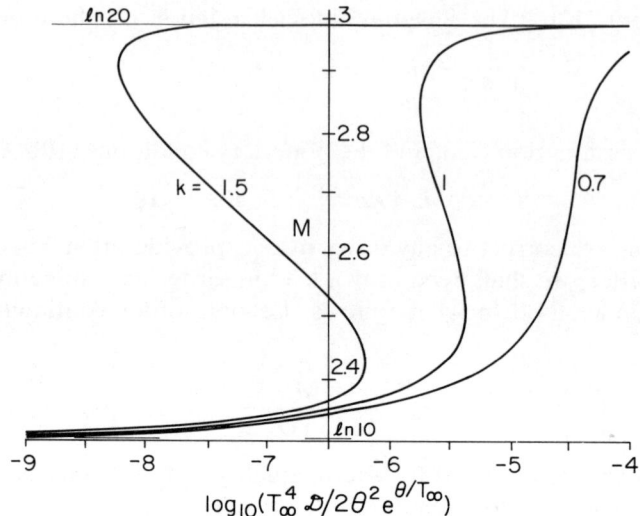

Fig. 9.2. *Steady-state responses when the combustion is nearly adiabatic. Drawn for* $L = 0.2$, $T_\infty = 0.2$, $X_\infty = 1$.

shape of an S appear to be associated with a flux of heat from the flame sheet to an environment that is hotter than the supply surface, the normal state of affairs in practice. However, it has never been proved that the inequalities (9) and (13b) will ensure an S-response in the limit $\theta \to \infty$.

3. General extinction and ignition analyses. The general extinction analysis under the conditions (9), (13b) follows that for the counterflow diffusion flame in § 8.3. On the whole upper branch of the S-response, the burning rate lies within $O(\theta^{-1})$ of the value (17), so that we write

$$M = M_e + \theta^{-1} M_1 + \cdots . \tag{33}$$

The combustion field on each side of the flame sheet is again frozen but identical to the equilibrium solution, here given by (16) with $Z_-(M_e) = T_s - L + Le^{M_e(1-1/r)}$, so far as leading terms are concerned. However, we now need

$$T_1 = M_1 L\left(1 - \frac{1}{r}\right) e^{M_e(1-1/r)} \quad \text{for } r < r_* \tag{34}$$

also, involving the perturbation M_1. It provides the stronger matching condition

$$\tilde{\phi} = -\tilde{\xi} - \tilde{M}_1 + \cdots \quad \text{as } \tilde{\xi} \to -\infty \quad \text{with } \tilde{M}_1 = \frac{(1+\tilde{\gamma})M_1}{T_*^2} \tag{35}$$

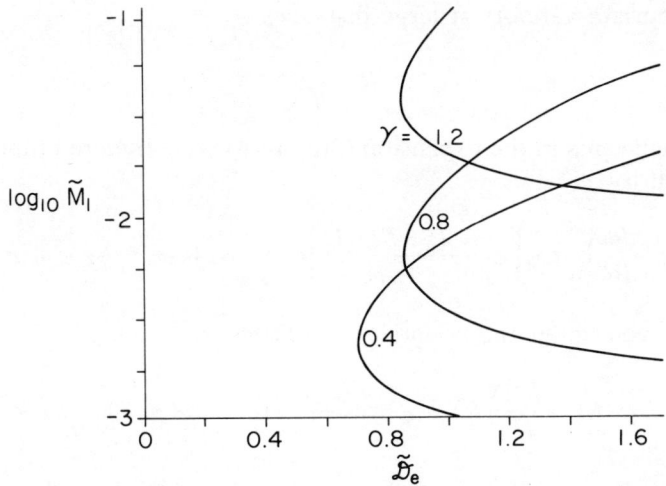

FIG. 9.3. *Extinction curves. For $\gamma \leq 0$ no turning point is found numerically.*

in the canonical problem (8.32), (8.33), at which the analysis finally arrives with

$$\tilde{\gamma} = 1 - \gamma, \qquad \tilde{\mathcal{D}}_e = \frac{r_*^4 T_*^8 \mathcal{D}}{4 M_e^2 \theta^3 e^{\theta/T_*}}; \tag{36}$$

here r_*, T_*, and γ have the definitions (18) and (19).

Since the problem is well posed under the weaker condition, it follows that

$$\tilde{M}_1 = -\lim_{\tilde{\xi} \to -\infty} (\tilde{\phi} + \tilde{\xi}) \tag{37}$$

can be calculated as a function of $\tilde{\mathcal{D}}_e$, thereby determining the response. This response is shown in Fig. 3 for various values of γ; it has been found numerically, but not yet been proved, that the curve turns to form a C whenever γ is positive, consistent with the conjecture that the inequality (9) must hold for an S-response.

The general ignition analysis that was promised in § 8.6 will now be presented. The starting point is the analog of the assumption (33), namely

$$M = M_w + \theta^{-1} M_1 + \cdots, \tag{38}$$

where M_w has the frozen value (12b). Correspondingly, the reaction all takes place far from the surface, the combustion field being frozen to all orders at any finite r. (The latter is ensured by the inequality (13b).) Thus, the formulas

$$T_0 = T_s - L + L e^{M_w(1 - 1/r)}, \qquad T_1 = M_1 L\left(1 - \frac{1}{r}\right) e^{M_w(1 - 1/r)} \tag{39}$$

are obtained by satisfying conditions (10b, c) at the surface; but these do not satisfy the condition (10d) at infinity.

The appropriate variable at large distances is

$$R = \frac{r}{\theta}, \qquad (40)$$

and then coefficients in the expansion (30) are to be considered functions of R. The equation for ϕ is

$$\frac{1}{R^2}\frac{d}{dR}\left(R^2 \frac{d\phi}{dR}\right) = \mathscr{D}_w\left[\phi + \frac{(T_a - T_\infty)M_w}{T_\infty^2 R}\right]e^{-\phi} \quad \text{with } \mathscr{D}_w = \tfrac{1}{2}\mathscr{D}X_\infty \theta^2 e^{-\theta/T_\infty}; \quad (41)$$

it is to be solved under the boundary conditions

$$\phi = Le^{M_w}\left(-M_1 + \frac{M_w}{R}\right)T_\infty^2 + \cdots \quad \text{as } R \to 0, \qquad \phi = o(1) \quad \text{as } R \to \infty, \quad (42)$$

the first of which comes from matching with the expansion coefficients (39) for finite r.

Since the problem is well posed under the weaker condition

$$\lim_{R \to 0} R\phi = Le^{M_w}\frac{M_w}{T_\infty^2}, \qquad (43)$$

it follows that

$$M_1 = \lim_{R \to 0}\left(\frac{M_w}{R} - \frac{T_\infty^2 \phi}{Le^{M_w}}\right) \qquad (44)$$

can be calculated as a function of $\mathscr{D}_w$, thereby determining the response. The

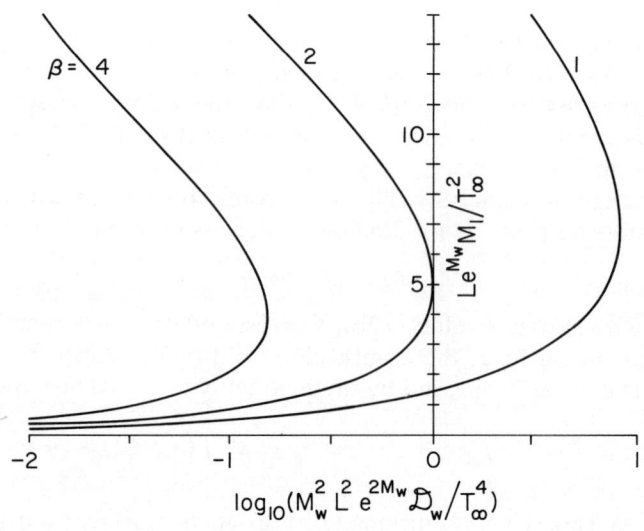

FIG. 9.4. Ignition curves. For $\beta \leq 0$ no turning point is found numerically.

relation between the scaled parameters $(Le^{M_w}/T_\infty^2)M_1$, $(M_w^2 L^2 e^{2M_w}/T_\infty^4)\mathcal{D}_w$ depends only on

$$\beta = \frac{T_a - T_\infty}{Le^{M_w}}; \qquad (45)$$

Fig. 4 shows the corresponding response for several positive values of β. It is found numerically, but has never been proved, that the curve does not turn for nonpositive values of β, consistent with the conjecture that the inequality (9) must hold for an S-response.

It is worth noting that $\mathcal{D}$ is $O(\theta^3 e^{\theta/(T_\infty + \gamma X_\infty)})$ at extinction compared to $O(\theta^{-2} e^{\theta/T_\infty})$ at ignition, so that E lies to the left of I, as required. The part between the ignition and extinction points, i.e. the middle branch of the S, will not be discussed here because it is very similar to that for the counterflow diffusion flame (§ 8.4.). Again the partial-burning and premixed flames occur, with the same structures for their reaction zones. Since the stability analysis of the branch is confined to the reaction zones, it is the same here as in § 8.5, so no further discussion is necessary.

4. Surface equilibrium. There are at least two ways in which the fuel can be supplied as a liquid at the surface. The sphere can be completely liquid, i.e. a fuel drop, or it can be a liquid-saturated porous solid. Whatever the method of supply, it is difficult to justify prescription of the surface temperature T_s unless we abandon the specification of L. Maintaining a value of T_s would, in general, require heating or cooling the liquid at the surface, and that would upset the heat balance represented by the boundary condition (10c). Prescription of surface temperature should, more realistically, be replaced by the requirement of liquid-vapor equilibrium at the surface, i.e. the Clausius–Clapeyron relation

$$Y_s = \left(\frac{T_s}{T_b}\right)^\beta e^{\hat{\theta}/T_b - \hat{\theta}/T_s} \quad \text{with } \beta, \hat{\theta} \text{ constants.} \qquad (46)$$

Here T_b is the boiling temperature, i.e. the value of T_s for which $Y_s = 1$; it is related to the pressure level in the combustion field by

$$T_b^\beta e^{-\hat{\theta}/T_b} = \frac{p_c}{k} \quad \text{with } k = \text{const.} \qquad (47)$$

In order to transform the Clausius–Clapeyron relation into a temperature condition, we eliminate Y_s between it and the Shvab–Zeldovich relation (7) to obtain

$$2\left(\frac{T_s}{T_b}\right)^\beta e^{\hat{\theta}/T_b - \hat{\theta}/T_s} = 2 - L + (T_\infty - T_s + L - 2)e^{-M}. \qquad (48)$$

This replaces the boundary condition (10b); clearly T_s changes with M.

Determination of the response $M(\mathcal{D})$ now requires that T_s be calculated afresh for each point. In general, iterations are involved because the structure problem determining M as a function of $\mathcal{D}$ contains T_s (see, for example, the definitions (36)). Different results are obtained according to the way in which $\mathcal{D}$ is varied, via p_c or the radius a of the sphere, the reason being that p_c appears in the boundary condition (48) through T_b. (Since ρ_r has been taken proportional to p_c, and M_r inversely proportional to the radius a of the sphere, the definition (2.23b) shows that $\mathcal{D}$ is proportional to $p_c^\gamma a^2$.)

Normandia and Ludford have considered the response when $\mathcal{D}$ is varied via a. In particular, they find that the curve is again S-shaped when the inequalities (9), (13b) are satisfied by the T_s corresponding to small $\mathcal{D}$, i.e. the solution of (48) when M has the value (12b). (There is always such a solution, satisfying the left inequality automatically.) The previous analysis (for T_s fixed) is qualitatively, but not quantitatively, correct: while extinction and ignition analyses can be performed as in § 3, Figs. 3 and 4 cannot be carried over because the variations in T_s, though only $O(\theta^{-1})$ on the upper and lower branches of the S-response, modify the extinction and ignition values. These modifications have been worked out by Normandia and Ludford.

The results are strikingly different when $\mathcal{D}$ is varied via p_c, as Janssen & Ludford (1983a) have shown. The S is replaced by the rather odd shapes in Fig. 5, which covers the practical cases of heat being conducted to the environment. These responses were so unexpected that, to obtain more confidence in their validity, numerical integrations of the problem (10a, c, d), (48) were performed for $\theta = 10$. In every instance, the numerical results confirmed the essential features of the asymptotic responses. Nothing similar has apparently been obtained in combustion theory, and certainly not in previous studies of diffusion flames.

Two features of these responses deserve to be pointed out, since they contradict conventional wisdom. First, the Burke–Schumann value (17) is not

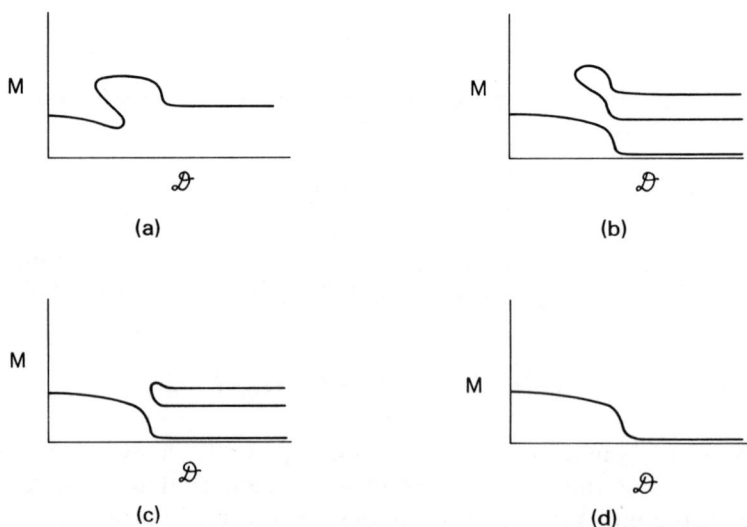

FIG. 9.5. *Sketch of steady state responses when there is surface equilibrium.*

attained on any of the curves as $\mathscr{D} \to \infty$. The Burke–Schumann solution discussed in § 1 is considered a good approximation since in practice $\mathscr{D}$ is large. But, instead of standing off from the sphere for high pressures, separating equilibrium regions, the flame sheet actually moves to the surface, forming the second (Buckmaster) equilibrium limit mentioned in connection with the condition (20). Secondly, the burning rate decreases over most of each response, whereas the general belief is that it should increase. Negative slope of the response curve is thought to be inevitably associated with instability, and indeed we found that to be the case for the S-response at the end of § 3. But experience refutes such a conclusion here: by and large, fuel drops do burn steadily.

A physically reasonable range of pressures focuses attention on the left ends of the curves in Fig. 5, and there the decrease of burning rate with pressure is by no means insignificant. Moreover, the computed response curve for methanol over the range 0.002 to 500 atmospheres clearly shows the steady decline (Janssen & Ludford (1983b)). Experiment should therefore be able to give a clear-cut decision on the physical reality of this unexpected phenomenon.

LECTURE 10

Free-Boundary Problems

Throughout these lectures we have ensured that the reaction terms vanish everywhere except in a thin (flame) sheet, whose location has to be found as part of the solution. So far this free boundary has been either a plane, a circular cylinder, a sphere, or a perturbation of one of these; we now consider problems with more complicated free boundaries.

There are four different ways of confining the reaction to a sheet.
  (i) Adopt the delta-function model discussed in Lecture 7.
  (ii) Take the hydrodynamic limit.
  (iii) Take the Burke–Schumann limit.
  (iv) Use activation-energy asymptotics.

In this lecture, which is an expanded version of Buckmaster (1983), we shall briefly mention examples of (ii) and (iii), but most of the discussion will deal with parabolic problems for premixed flames that arise from (iv).

1. The hydrodynamic limit. The jump conditions (3.9), (3.10) imply that the flow will be diffracted by an inclined flame. If the flame speed W is much smaller than the speed U of the fresh gas, i.e.

$$W = \varepsilon U \quad \text{with } \varepsilon \ll 1, \tag{1}$$

a uniform flow

$$\mathbf{v}^i = (U, 0) \tag{2}$$

is turned by a plane flame so that the velocity of the burnt gas is

$$\mathbf{v}^e = (U, \pm \varepsilon U(\sigma - 1)) + o(\varepsilon, \varepsilon), \tag{3}$$

where the sign is opposite to that of the flame slope (Fig. 1). The corresponding jump in pressure across the flame is $O(\varepsilon)$. Such a deflection of the streamlines is a fundamental characteristic of tube flames (Fig. 2), which are often slender (i.e. correspond to small ε).

In order to describe the shape of the flame when the prescribed efflux from the tube is the plane flow $U(f(\eta), 0)$, we look for i(nterior) and e(xterior) solutions

$$u = \begin{cases} Uf(\eta) + \varepsilon u_1^i(\varepsilon\chi, \eta) + \cdots, \\ Uf(\eta) + \varepsilon u_1^e(\varepsilon\chi, \varepsilon\eta) + \cdots, \end{cases} \quad v = \begin{cases} \varepsilon^2 v_2^i(\varepsilon\chi, \eta) + \cdots, \\ \varepsilon v_1^e(\varepsilon\chi, \varepsilon\eta) + \cdots. \end{cases} \tag{4}$$

These expansions are consistent with the slender-flame approximation, Euler's equations and the jump conditions. Both the flame shape and the flow field can be constructed in a straightforward manner (Buckmaster & Crowley (1983)).

In fact, a differential equation for the shape

$$\eta = \pm F(\varepsilon\chi) \tag{5}$$

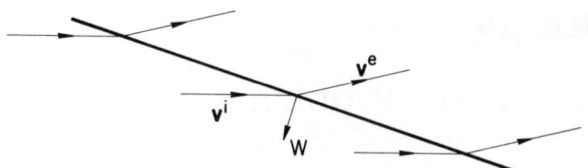

FIG. 10.1. *Diffraction of streamlines at flame surface.*

of the flame can be deduced immediately. Since v^i is $O(\varepsilon^2)$, it does not contribute to the $O(\varepsilon)$ normal velocity ahead of the flame, so that the flame speed has the required value (1) if

$$1 + f(F)F' = 0, \tag{6}$$

a result that is also true for axisymmetric flow. When

$$f(\eta) = 1 - \frac{\eta^2}{a^2} \quad \text{with } |\eta| \le a, \tag{7}$$

i.e. the flow is Poiseuille, the flame shape is given by

$$F - \frac{F^3}{3a^2} = -\varepsilon \chi. \tag{8}$$

Only for slender flames can the shape be determined directly from the given flow leaving the burner.

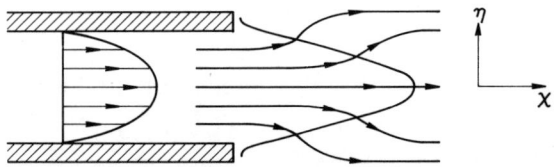

FIG. 10.2. *Tube flame.*

2. The Burke–Schumann limit. This limit arises in the context of diffusion flames as $\mathcal{D} \to \infty$. As we saw in § 8.1, the fuel is then absent on one side of the flame sheet and the oxidant on the other. The problem considered by Burke and Schumann in their seminal work was the divided flow of fuel and oxidant through concentric tubes with the flame attached to the rim of the inner tube (Fig. 3). If the oxidant is in excess, the flame terminates at the axis and is said to be overventilated; if there is an excess of fuel, termination occurs at the outer tube, an underventilated condition.

Burke and Schumann neglected the effect of the flame on the flow, i.e. they adopted the constant-density approximation over 50 years ago. It is curious then, in view of the enormous insight provided by their well-known analysis, that to this day one finds, at scientific forums, objections to the approximation

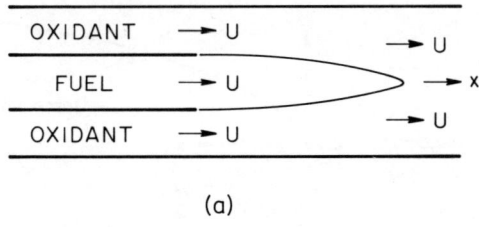

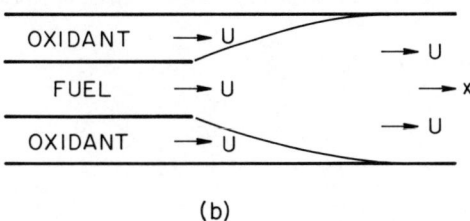

FIG. 10.3. *Burke–Schumann's problem: diffusion flame* (a) *overventilated,* (b) *underventilated.*

when it is used in activation-energy asymptotics. They also neglected longitudinal diffusion, a step that can be legitimized by setting

$$x = \frac{x}{U} \tag{9}$$

and letting $U \to \infty$. The equations to be considered are then

$$\frac{\partial T}{\partial x} = \frac{\partial^2 T}{\partial y^2}, \quad \frac{\partial X}{\partial x} = \mathcal{H}^{-1}\frac{\partial^2 X}{\partial y^2}, \quad \frac{\partial Y}{\partial x} = \mathcal{L}^{-1}\frac{\partial^2 Y}{\partial y^2} \tag{10}$$

under the jump conditions

$$\delta(T) = \delta(X) = \delta(Y) = \delta\left(\frac{\partial T}{\partial y} + 2\mathcal{H}^{-1}\frac{\partial X}{\partial y}\right) = \delta\left(\frac{\partial T}{\partial y} + 2\mathcal{L}^{-1}\frac{\partial Y}{\partial y}\right) = 0. \tag{11}$$

The plane version of the equations used by Burke and Schumann has been written because we shall consider a simpler (but related) problem than theirs. General Lewis numbers have also been introduced, which necessitates a slight modification of the jump conditions (9.21), (9.22).

Consider a plate, coincident with the negative x-axis, separating parallel and equally fast flows of oxidant and fuel (Fig. 4). The initial conditions are

$$T = \begin{cases} T_{f+}, \\ T_{f-}, \end{cases} \quad X = \begin{cases} X_f, \\ 0, \end{cases} \quad Y = \begin{cases} 0 \\ Y_f \end{cases} \quad \text{for } y \gtrless 0 \text{ at } x = 0; \tag{12}$$

in fact, X is required to vanish everywhere below the flame sheet and Y everywhere above. The solution of the problem (10)–(11) can be written in

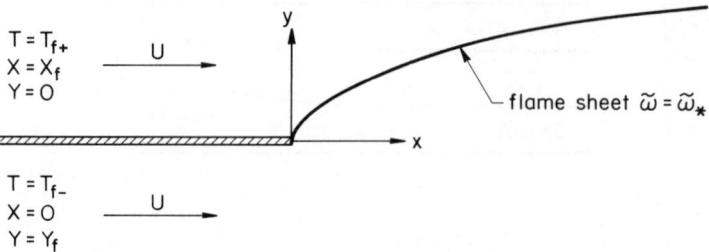

Fig. 10.4. *Simpler version of Burke–Schumann problem.*

terms of the similarity variable

$$\tilde{\omega} = \frac{y}{x^{1/2}} \tag{13}$$

as

$$X = \begin{cases} X_f\left[1 - \dfrac{\text{erfc}\,(\mathcal{H}^{1/2}\tilde{\omega}/2)}{\text{erfc}\,(\mathcal{H}^{1/2}\tilde{\omega}_*/2)}\right], \\ 0, \end{cases} \quad Y = \begin{cases} 0 \\ Y_f\left[1 - \dfrac{\text{erfc}\,(\mathcal{L}^{1/2}\tilde{\omega}/2)}{\text{erfc}\,(\mathcal{L}^{1/2}\tilde{\omega}_*/2)}\right] \end{cases} \quad \text{for } \tilde{\omega} \gtrless \tilde{\omega}_*, \tag{14}$$

where $\tilde{\omega}_*$ is given by

$$X_f \mathcal{L}^{1/2} e^{-\mathcal{H}\tilde{\omega}_*^2/4} \text{erfc}\,(-\mathcal{L}^{1/2}\omega_*/2) = Y_f \mathcal{H}^{1/2} e^{-\mathcal{L}\tilde{\omega}_*^2/4} \text{erfc}\,(\mathcal{H}^{1/2}\omega_*/2). \tag{15}$$

Corresponding formulas for T could also be written.

When the reactants are supplied in stoichiometric proportions (i.e. $X_f = Y_f$) and when in addition $\mathcal{H} = \mathcal{L}$, the flame sheet coincides with the positive x-axis, i.e. $\tilde{\omega}_* = 0$. A decrease (increase) in X_f or an increase (decrease) in $\mathcal{H}$ moves the flame sheet up (down).

Our simple problem has yielded to analytical treatment, but in general the Burke–Schumann limit involves numerical integration of a type that arises in the free-boundary problems of activation-energy asymptotics to be treated next.

3. NEF tips. The most interesting examples of free-boundary problems uncovered by activation-energy asymptotics concern NEFs, and discussions of these examples have almost invariably adopted the constant-density approximation. The relevant equations and jump conditions have been developed in § 4.4.

The starting point is a steady plane deflagration in a uniform flow with speed U greater than 1 (Fig. 5). The solution is

$$T = \begin{cases} T_f + Y_f e^n, \\ H_f \equiv T_b, \end{cases} \quad h = \begin{cases} -lY_f n e^n \\ 0 \end{cases} \quad \text{for } n \lessgtr 0, \tag{16}$$

formulas that have already been used in connection with plane-flame stability

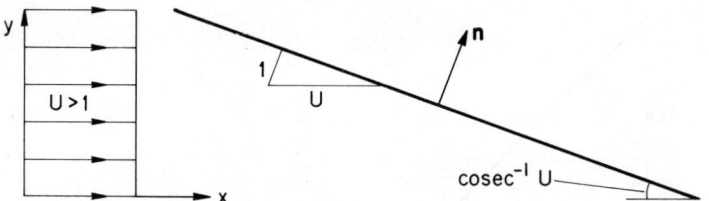

FIG. 10.5. Plane flame in uniform flow.

(§ 5.2). Now consider the effect of introducing an adiabatic, noncatalytic wall at the plane $y=0$. For simplicity we shall take $l=0$, so that h vanishes identically and the problem reduces to one for T alone, namely the free-boundary problem

$$U\frac{\partial T}{\partial x} = \nabla^2 T \quad \text{for } n<0 \text{ and } y>0, \tag{17}$$

$$T \to T_b, \quad \frac{\partial T}{\partial n} \to Y_f \quad \text{as } n \to 0-, \tag{18}$$

$$\frac{\partial T}{\partial y} = 0 \quad \text{at } y=0, \tag{19}$$

$$T \to T_f + Y_f e^n \quad \text{as } x \to -\infty \text{ with } n \text{ fixed.} \tag{20}$$

The last condition comes from the requirement that, far from the wall, the temperature field has the description (16a); this structure gives way to a two-dimensional combustion field as the wall is approached. Here we have the problem of a plane-flame tip for $\mathscr{L}=1$, the wall corresponding to the line of symmetry. No attempt has been made to solve it in this form because of its elliptic nature.

If U is large, the formulation becomes parabolic, giving a classical problem of Stefan type. Thus, introducing the coordinate (9) and letting $U \to \infty$ converts the problem (17)–(20) into

$$\frac{\partial T}{\partial \chi} = \frac{\partial^2 T}{\partial y^2} \quad \text{for } 0<y<F(\chi), \tag{21}$$

$$T \to T_b, \quad \frac{\partial T}{\partial y} \to Y_f \quad \text{as } y \to F(\chi)-0, \tag{22}$$

$$\frac{\partial T}{\partial y} = 0 \quad \text{at } y=0, \tag{23}$$

$$T \to T_f + Y_f e^{y+\chi} \quad \text{as } \chi \to -\infty, \tag{24}$$

where

$$y = F(\chi) \tag{25}$$

is the free boundary and the origin has been taken at the intersection of the

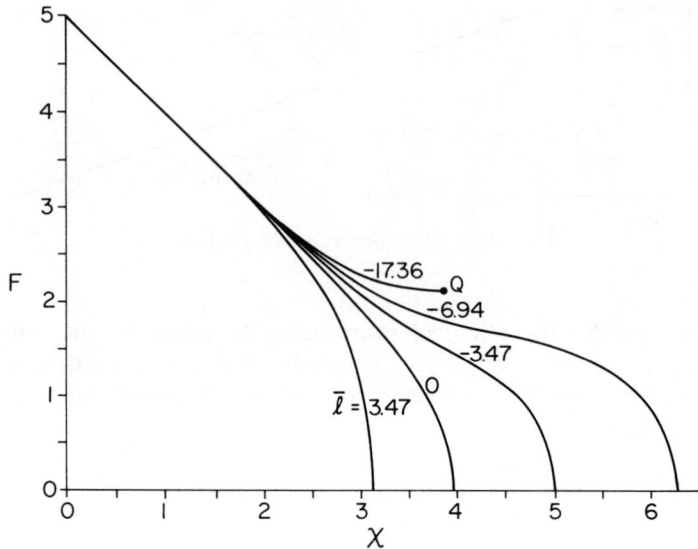

FIG. 10.6. Plane NEF tips.

undisturbed flame with the wall, i.e.

$$\lim_{\chi \to -\infty} [F(\chi) + \chi] = 0. \tag{26}$$

This problem is identical to that of a plane flame approaching a parallel adiabatic wall, with χ playing the role of time.

The solution of the Stefan problem (21)–(24), and similar ones that are discussed later, must be obtained numerically. Meyer's (1977) method of lines is particularly well suited to this task and is outlined in an appendix. It was used to obtain Fig. 6, which reveals a significant increase in flame speed as the flame sheet approaches the axis, i.e. near the tip. This adjustment on the diffusion-length scale of a NEF smooths out the sharp tip predicted by the hydrodynamic analysis in § 1.

A similarity solution can be constructed for such problems in the neighborhood of the tip. Setting

$$T = T_b + (\chi_t - \chi)^{1/2} G(\tilde{\omega}) \quad \text{with } \tilde{\omega} = (\chi_t - \chi)^{-1/2} y, \tag{27}$$

where χ_t defines the location of the tip, leads to the differential equation

$$G'' - \tfrac{1}{2}\tilde{\omega}G' + \tfrac{1}{2}G = 0 \tag{28}$$

and the boundary conditions

$$G'(0) = 0, \quad G(\omega_*) = 0, \quad G'(\omega_*) = Y_f. \tag{29}$$

Here the constant $\tilde{\omega}_*$, which has to be found, determines the locally parabolic flame sheet. The solution of (28) satisfying the conditions (29b, c) is

$$G = Y_f \tilde{\omega}_* e^{-\tilde{\omega}_*^2/4} \tilde{\omega} \int_{\tilde{\omega}_*}^{\tilde{\omega}} \frac{e^{x^2/4}}{x^2} dx; \tag{30}$$

the condition (29a) is then satisfied if

$$\tilde{\omega}_* e^{-\tilde{\omega}_*^2/4} \int_0^{\tilde{\omega}_*/2} e^{x^2} dx = 1, \quad \text{i.e. } \omega_* = 1.85. \tag{31}$$

The similarity solution breaks down in the neighborhood of the tip where the flame slope is $O(1)$, but the extent of this neighborhood is given by

$$\chi_t - \chi = O(U^{-2}) \tag{32}$$

and, hence, can be made arbitrarily small by taking U large enough.

For $l \neq 0$, i.e. $\mathcal{L} \neq 1$, the function h no longer vanishes and the problem (21)–(24) must be augmented with equations and conditions for h, in particular

$$\frac{\partial h}{\partial y} = 0 \quad \text{for } y = 0. \tag{33}$$

The effects of Lewis number are also shown in Fig. 6. Decreasing from 1 causes the tip to elongate and eventually assume a bulbous form. It is tempting to terminate solutions of the latter type at the point Q in Fig. 6, for they then closely resemble the open flame tips seen in the combustion of lean hydrogen or rich heavy-hydrocarbon mixtures, mixtures for which the effective Lewis number is significantly less than 1. But the mathematics gives no clear-cut reason for doing this; the only suggestion is the marked decrease in temperature along the flared portion of the flame.

The negative flame speeds associated with those portions of the flame with positive slope, although curious, do not violate the physics. There is a diffusive flux of reactant in the y-direction towards the flame sheet to maintain the combustion.

4. NEF wall-quenching. When the wall is not adiabatic, the problem is no longer relevant to flame tips. For a cooled wall it models, in a rough sense, the behavior of a flame near a burner rim (Fig. 7). It corresponds more precisely to the propagation of a plane wave towards a parallel cooled wall, but our discussion will be couched in terms of burner flames.

A flame can be stabilized on a Bunsen burner only under a limited range of conditions. Outside this range either "blow-off" will occur so that the flame becomes detached from the burner, or "flashback" will take place, the flame traveling to the base of the burner via the inner wall of the tube. The tendency for flashback is easy to understand: a premixed flame will travel upstream unless the gas speed is greater than the flame speed, and at the surface of the burner tube the gas speed falls to zero. Flashback is, therefore, inevitable unless there is some mechanism to prevent the flame from reaching the surface.

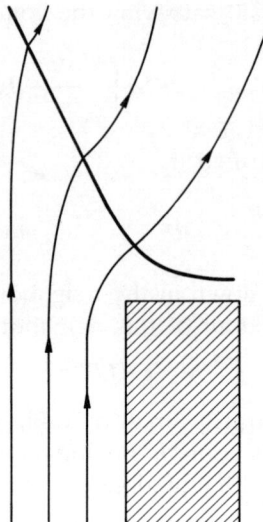

FIG. 10.7. *Flame near a burner rim.*

Heat transfer from the gas to the tube, quenching the reaction that sustains the flame, is commonly regarded as one such mechanism; the free-boundary problem under discussion illustrates this mechanism.

The boundary conditions at the wall are taken to be

$$\frac{\partial T}{\partial y} = \frac{k(T - T_f)}{\theta}, \quad \frac{\partial Y}{\partial y} = 0 \quad \text{at } y = 0, \tag{34}$$

i.e.

$$\frac{\partial T}{\partial y} = 0, \quad \frac{\partial h}{\partial y} = k(T - T_f) \quad \text{at } y = 0 \tag{35}$$

in the NEF formulation. At the same time, the problem will be generalized slightly by considering a not-necessarily-uniform flow

$$\mathbf{v} = (Uf(y), 0) \quad \text{as } U \to \infty. \tag{36}$$

This is more realistic than a uniform flow when the no-slip condition at the wall is satisfied by taking $f(0) = 0$. We shall set $l = 0$ so as to filter out the enthalpy loss or gain at the flame due to unbalanced diffusion of temperature and reactant.

The problem is again numerical, but certain features can be derived analytically. So long as the flame intersects the wall, a similarity solution describes its behavior there; when it extends to $\chi = \infty$, there is an asymptotic solution.

Consider first

$$f(y) \equiv 1; \tag{37}$$

later we will briefly discuss the more realistic choice (55). We shall assume that

$$T \to T_b \quad \text{as } \chi \to \infty, \tag{38}$$

the approach being exponential in χ, and that
$$\chi^{-1/2} F(\chi) \to 0 \quad \text{as } \chi \to \infty, \tag{39}$$
where F is the free-boundary function (25). Self-consistency of the resulting asymptotic description will be checked in due course.

We start with the problem for h (which has no jumps at the flame sheet) and write
$$h = h^{(1)} + h^{(2)} + h^{(3)}, \tag{40}$$
where $h^{(1)}$, $h^{(2)}$, $h^{(3)}$ satisfy the heat equation (21) individually, satisfy the boundary conditions
$$\frac{\partial h^{(1)}}{\partial y} = kY_f, \quad \frac{\partial h^{(2)}}{\partial y} = 0, \quad \frac{\partial h^{(3)}}{\partial y} = k(T - T_f - Y_f) \quad \text{at } y = 0, \tag{41}$$
and vanish exponentially rapidly as $y \to \infty$. The condition (38) ensures that $h^{(3)}$ is exponentially small as $\chi \to \infty$ and so need not be considered further. The asymptotic behaviors of the two remaining functions are determined by similarity solutions of the heat equation. Thus
$$h^{(1)} \sim \chi^{1/2} G(\tilde{\omega}) \quad \text{with } \tilde{\omega} = \frac{y}{\chi^{1/2}}, \quad G = kY_f\left(\tilde{\omega}\,\text{erfc}\,\frac{\tilde{\omega}}{2} - \frac{2}{\sqrt{\pi}} e^{-\tilde{\omega}^2/4}\right) \tag{42}$$
and $h^{(2)}$ is asymptotically a sum of the eigensolutions
$$\chi^{-1/2-n} E(\tilde{\omega}) \quad \text{with } E = e^{-\tilde{\omega}^2/8} D_{2n}\!\left(\frac{\tilde{\omega}}{\sqrt{2}}\right); \tag{43}$$
here n is a nonnegative integer and D is the parabolic cylinder function. Since all the eigensolutions vanish as $\chi \to \infty$, we need not consider $h^{(2)}$ further.

Only $h^{(1)}$ is left and from it we find
$$\frac{h}{\chi^{1/2}} = -\frac{2kY_f}{\sqrt{\pi}} + O(\tilde{\omega}) \quad \text{for } \tilde{\omega} \text{ small.} \tag{44}$$
In view of the hypothesis (39), setting $\tilde{\omega} = \chi^{-1/2} F$ in this expansion gives the value of h at the flame sheet, so that
$$\phi_* = \left(\frac{2kY_f}{\sqrt{\pi}T_b^2}\right)\chi^{1/2}[1 + o(1)] \quad \text{as } \chi \to \infty. \tag{45}$$
Consider now the asymptotic behavior of
$$T = T_b + Y_f \bar{T}. \tag{46}$$
The appropriate modification of the problem (21)–(24) is
$$\frac{\partial \bar{T}}{\partial \chi} = \frac{\partial^2 \bar{T}}{\partial y^2} \quad \text{for } 0 < y < F(\chi), \tag{47}$$
$$\bar{T} \to 0, \quad \frac{\partial \bar{T}}{\partial y} \to \exp\left(-\frac{\phi_*}{2}\right) \quad \text{as } y \to F(\chi) - 0, \tag{48}$$
$$\frac{\partial \bar{T}}{\partial y} = 0 \quad \text{at } y = 0; \tag{49}$$

the initial conditions (at $\chi = -\infty$) are omitted. The solution is

$$F(\chi) \sim \tilde{\omega}_* \chi^{1/4}, \quad \bar{T} \sim \exp\left(-\frac{\phi_*}{2}\right)\chi^{1/4}G(\tilde{\omega}) \quad \text{with } \tilde{\omega} = y/\chi^{1/4}, \qquad (50)$$

where

$$G'' + a^2 G = 0, \quad G'(0) = 0, \quad G(\tilde{\omega}_*) = 0, \quad G'(\tilde{\omega}_*) = 1 \quad \text{with } a^2 = \frac{kY_f}{2\sqrt{\pi}T_b^2}. \qquad (51)$$

Note that the result (50a) is consistent with the assumption (39) under which it was obtained. The constant $\tilde{\omega}_*$, which has to be found, determines the asymptotic shape (50a) of the flame sheet. The solution of the equation (51a) satisfying the conditions (51b, d) is

$$G = -\frac{\cos a\tilde{\omega}}{a \sin a\tilde{\omega}_*}; \qquad (52)$$

the condition (51c) is then satisfied if

$$a\tilde{\omega}_* = (n + \tfrac{1}{2})\pi \qquad (53)$$

for some integer n. A further requirement comes from the maximum principle for the heat equation, namely $T \le T_b$ everywhere. The only acceptable n is zero and we have

$$\tilde{\omega}_* = \frac{\pi^{5/4} T_b}{\sqrt{2kY_f}}. \qquad (54)$$

A similar analysis is possible when

$$f(y) \equiv y, \qquad (55)$$

a more realistic choice physically. Now

$$h^{(1)} \sim \chi^{1/3} G(\tilde{\omega}) \quad \text{with } \tilde{\omega} = \frac{y}{\chi^{1/3}}, \quad G = \frac{kY_f \tilde{\omega}\Gamma(-1/3, \tilde{\omega}^3/9)}{\Gamma(-1/3)}, \qquad (56)$$

so that

$$\frac{h}{\chi^{1/3}} = -\frac{3^{2/3} kY_f}{\Gamma(2/3)} + O(\tilde{\omega}) \quad \text{for } \tilde{\omega} \text{ small}; \qquad (57)$$

and

$$\phi_* = \frac{3^{2/3} kY_f T_b^2}{\Gamma(2/3)} \chi^{1/3}[1 + o(1)] \quad \text{as } \chi \to \infty \qquad (58)$$

provided

$$\chi^{-1/3} F(\chi) \to 0 \quad \text{as } \chi \to \infty. \qquad (59)$$

The $\bar{T}$-problem has the solution

$$F(\chi) \sim \tilde{\omega}_* \chi^{2/9}, \quad \bar{T} \sim \exp\left(-\frac{\phi_*}{2}\right)\chi^{2/9} G(\tilde{\omega}) \quad \text{with } \tilde{\omega} = \frac{y}{\chi^{2/9}}, \qquad (60)$$

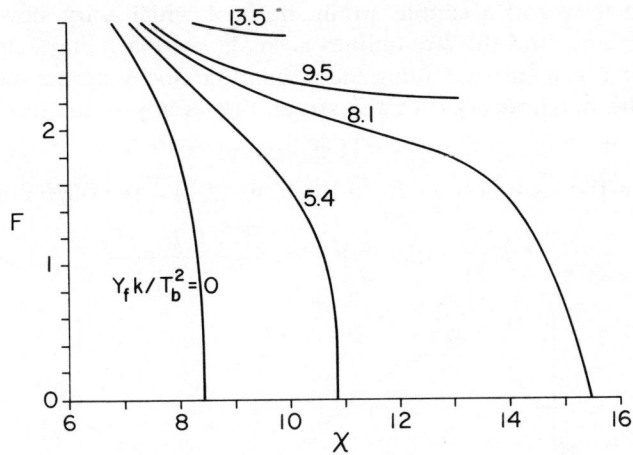

FIG. 10.8. *Flame-sheet profiles for cold-wall problem.*

where

$$G'' + a^2 \tilde{\omega} G = 0, \quad G'(0) = 0, \quad G(\tilde{\omega}_*) = 0, \quad G'(\omega_*) = 1$$

$$\text{with } a^2 = 3^{-1/3} \frac{kY_f}{2\Gamma(2/3)T_b^2}. \tag{61}$$

Finally, the maximum principle leads us, as before, to the conclusion that

$$G = -\frac{\tilde{\omega}^{1/2} J_{-1/3}(2a\tilde{\omega}^{3/2}/3)}{a\tilde{\omega}_* J_{2/3}(2a\tilde{\omega}_*^{3/2}/3)}, \quad \omega_* = \left(\frac{9r^2}{4a^2}\right)^{1/3}, \tag{62}$$

where r is the smallest zero of $J_{-1/3}(r)$.

These results suggest a simple dichotomy: for sufficiently small heat loss, the flame eventually intersects the wall (a necessary condition for flashback); but for larger values of k the flame ultimately moves away from the wall, in particular according to the result (60a) for the flow (55), and flashback cannot occur. Figure 8 shows Buckmaster's numerical solutions; clearly an increase in k tends to reduce the flame speed near the wall. For moderate values of k, he obtained slightly bulbous shapes (cf. flame tips for $\mathscr{L} < 1$), suggesting that this would be the case for all sufficiently large k. Numerical re-examination of the question has, however, provided good evidence that the dichotomy mentioned above is real.

5. Straining NEFs.

So far we have dealt only with parallel flows, but parabolic free-boundary problems can also be formulated for the more general velocity fields

$$\mathbf{v} = U\mathbf{q}\left(\frac{x}{U}, \frac{y}{U}\right) \quad \text{with } U \gg 1. \tag{63}$$

These are fast flows with $\nabla \mathbf{v} = O(1)$; more precisely, each is effectively made up of a uniform flow and a simple strain, both of which vary slowly. The angle between the flame and the streamlines is small, so that it stays close to a single streamline. If χ is measured along the streamline and y perpendicular to it, the velocity in the neighborhood of the streamline is approximated by

$$\mathbf{v} = (Uq_0, -q_0'y) \tag{64}$$

where $q_0(\chi)$ is the speed on $y = 0$. In the limit $U \to \infty$ the NEF equations (4.30) become

$$\left(q_0 \frac{\partial}{\partial \chi} - q_0' y \frac{\partial}{\partial y}\right)(T, h) = \frac{\partial^2(T, h + lT)}{\partial y^2}, \tag{65}$$

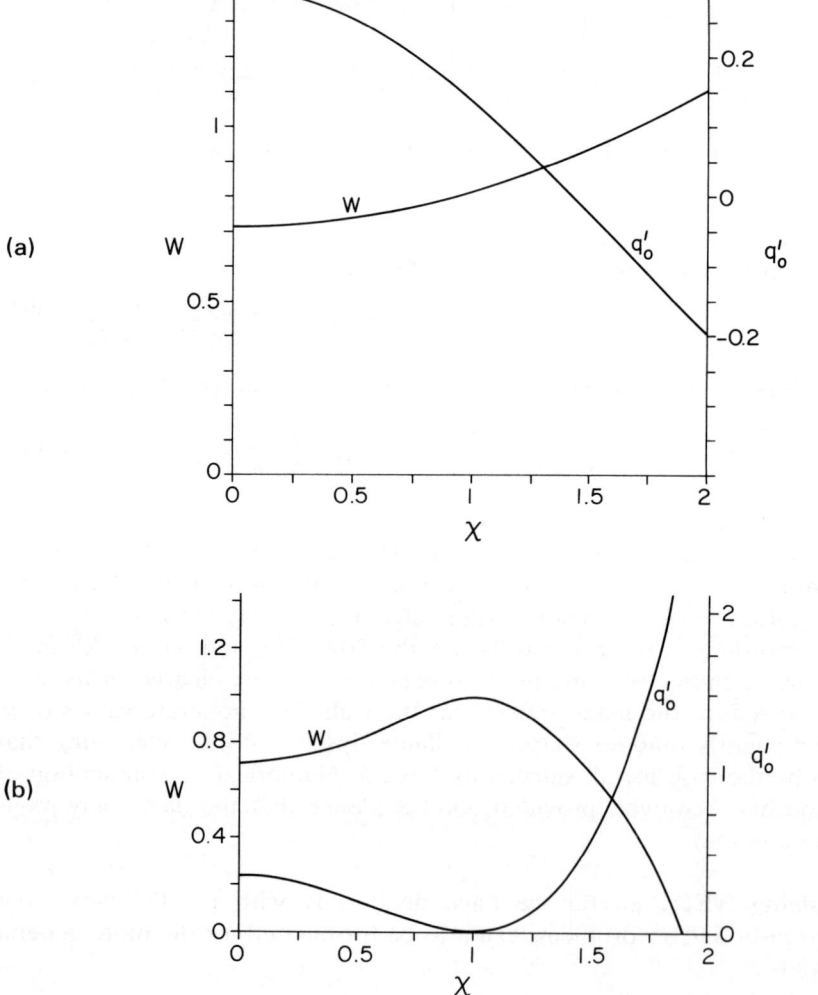

FIG. 10.9. Variation of flame speed W and stretch q_0' with χ for the more general nonuniform velocity fields (66a, b).

and these are amenable to numerical treatment for any choice of q_0. Modest variations in l do not have a substantial qualitative effect, so that we shall set $l = 0$.

Two examples will be considered, namely

$$q_0(\chi) = 0.35 \sin \chi \quad \text{for } 0 < \chi < \pi, \quad 0.35\left(\chi - \frac{2\chi^3}{3} + \frac{\chi^5}{5}\right) \quad \text{for } \chi > 0. \quad (66)$$

Each of these has a stagnation point as $\chi \to 0$, where the analysis leading to (65) breaks down; nevertheless the corresponding stagnation-point solution constructed in § 4.5 provides the correct initial conditions. Figure 9 shows, for each example, a correlation between the flame speed W and the strain rate q_0', a meaure of the stretch experienced by an element of the flame: as the stretch increases (decreases) the flame speed decreases (increases).

The conclusion is not necessarily valid in other circumstances, however, as can be seen from analytical results obtained by Buckmaster (1982) for a general, small but rapid velocity change

$$q_0 = 1 + \varepsilon Q(\bar{\chi}) \quad \text{with } 0 < \varepsilon \ll 1, \quad \bar{\chi} = \frac{\chi}{\varepsilon^2} \quad (67)$$

on the scale of χ. The wave speed has the explicit representation

$$W(\bar{\chi}) = 1 - \frac{1}{\sqrt{\pi}} \int_{-\infty}^{\bar{\chi}} \frac{Q'(\mu)}{\sqrt{\bar{\chi} - \mu}} d\mu, \quad (68)$$

a formula that, surprisingly enough, is valid for all values of l. There is no strong connection between the behavior of W and local changes in the stretch $\varepsilon Q'$.

The result (68) shows that Q can be chosen (with a degree of arbitrariness) to make W vanish at any given point. The possibility of decreasing the flame speed to zero at a point by adjusting the velocity field has not been demonstrated before.

6. Shearing NEFs.

We conclude our discussion of free-boundary problems by briefly describing the effect of shear on a premixed flame. The wall is removed and a linear shear flow is inserted for $y < 0$, the flow being assumed uniform for $y > 0$. More precisely,

$$f(y) = \begin{cases} 1 \\ 1 - \omega y \end{cases} \quad \text{for } y \gtrless 0 \quad (69)$$

in the expression (36).

Numerical results have been obtained, in particular for $\omega = 5$. Here we shall be content to describe the solution in the limit of very strong shear, i.e. $\omega \to \infty$. Then the equations in the lower half-plane simplify to

$$\frac{\partial T}{\partial x} = \frac{\partial h}{\partial x} = 0, \quad (70)$$

giving the solution

$$T = T_f, \quad h = 0; \tag{71}$$

these become boundary conditions (at $y = 0$) on the solution in the upper half-plane.

The asymptotic behavior as $\chi \to \infty$ is particularly simple, depending only on y:

$$T = \begin{cases} \dfrac{T_f + Y_f y}{F(\infty)}, \\ T_b, \end{cases} \quad h = \begin{cases} -\dfrac{lY_f y}{F(\infty)} & \text{for } y \leq F(\infty), \\ -lY_f \end{cases} \tag{72}$$

where

$$F(\infty) = e^T. \tag{73}$$

Numerical integration shows how the remote combustion field (16) is transformed into the asymptotic field (72). The flame speed vanishes as $\chi \to \infty$ and is negative or positive in the neighborhood of infinity accordingly as l is positive or negative.

Appendix. The method of lines. Consider the parabolic free-boundary problem (21)–(24). Let

$$T_n(y) = T(\chi_n, y), \quad F_n = F(\chi_n), \tag{74}$$

and approximate the partial differential equation (21) by the ordinary differential equation

$$T_n'' = \frac{T_n}{\Delta \chi} - \frac{T_{n-1}}{\Delta \chi} \quad \text{with } \Delta \chi = \chi_n - \chi_{n-1}. \tag{75}$$

Having determined T_{n-1} at the previous step, we must integrate this second-order equation for T_n along the line $\chi = \chi_n$ subject to three boundary conditions (22), (23). The integration will, therefore, determine not only T_n but also F_n.

Let $A(y)$ be the solution of (75a) with $A(0) = 1$, $A'(0) = 0$, and let $B(y)$ be the solution with $B(0) = B'(0) = 0$. Then

$$T_n = (A - B)b + B \tag{76}$$

describes a one-parameter family of solutions satisfying

$$T_n(0) = b, \quad T_n'(0) = 0, \tag{77}$$

and from it we may calculate

$$T_n' = (A' - B')b + B'. \tag{78}$$

Elimination of b from (76), (78) gives a relation of the form

$$T_n' = PT_n + Q, \tag{79}$$

where $P(y)$ and $Q(y)$ could be written in terms of A and B. The initial conditions (77) show that

$$P(0) = Q(0) = 0, \tag{80}$$

since P, Q (like A, B) are independent of the parameter b. The key step is to find differential equations for P, Q.

If T'_n and T''_n are eliminated between (75a) and (79), we are left with

$$T_n\left(P' + P^2 - \frac{1}{\Delta\chi}\right) + \left(Q' + PQ + \frac{T_{n-1}}{\Delta\chi}\right) = 0. \tag{81}$$

Since only T_n depends on b, the parentheses must separately vanish, giving a pair of first-order differential equations for P, Q. Numerical integration, under the initial conditions (80), then determines these functions.

The position $y = F_n$ of the free boundary can now be determined as a root of the equation

$$Y_f = P(F_n)T_b + Q(F_n), \tag{82}$$

obtained by substituting the flame-sheet conditions (22) in the relation (79) which is satisfied for all the functions T_n in the family, including the one we are seeking. The latter is then determined by integrating (79) (considered as a first-order differential equation for T_n) towards $y = 0$ with the initial condition $T_n(F_n) = T_b$.

The method is quick and efficient, and requires no iterations.

References

BUCKMASTER, J. D. (1982), Two examples of a stretched flame, *Quarterly Journal of Mechanics and Applied Mathematics*, 35, pp. 249–63.

BUCKMASTER, J. D. (1983a), Free boundary problems in combustion, in *Free Boundary Problems: Theory and Applications II*, A. Fasano and M. Primicerio, eds., Research Notes in Mathematics 79, Pitman, Boston.

BUCKMASTER, J. (1983b), Stability of the porous plug burner flame, SIAM *Journal on Applied Mathematics*, 43 (1983), pp. 1335–1349.

BUCKMASTER, J. (1984), Polyhedral flames—an exercise in bimodal bifurcation analysis, SIAM *Journal on Applied Mathematics* (in press).

BUCKMASTER, J. & CROWLEY, A. (1983), The fluid mechanics of flame tips, *Journal of Fluid Mechanics* (in press).

BUCKMASTER, J. & LUDFORD, G. S. S. (1982), *Theory of Laminar Flames*, University Press, Cambridge.

BUCKMASTER, J. & MIKOLAITIS, D. (1982), A flammability-limit model for upward propagation through lean methane/air mixtures in a standard flammability tube, *Combustion and Flame*, 45, pp. 109–119.

BUCKMASTER, J. D., NACHMAN, A., & TALIAFERRO, S. (1983), The fast-time instability of diffusion flames. *Physica D* (in press).

DANESHYAR, H., LUDFORD, G. S. S., & MENDES-LOPES, J. M. C. (1983), Effect of strain fields on burning rate, in *Nineteenth International Symposium on Combustion*, The Combustion Institute, Pittsburgh, pp. 413–421.

DANESHYAR, H., LUDFORD, G. S. S., MENDES-LOPES, J. M. C., & TROMANS, P. S. (1983), The influence of straining on a premixed flame and its relevance to combustion in SI engines, in *International Conference on Combustion in Engineering*, 1, Mechanical Engineering Publications, London, pp. 191–199.

GOLOVICHEV, V. I., GRISHIN, A. M., AGRANAT, V. M., & BERTSUN, V. N. (1978), Thermokinetic oscillations in distributed homogeneous systems, *Doklady Akademii Nauk SSSR*, 241, pp. 305–308 = *Soviet Mathematics Doklady*, 23, pp. 477–479.

JANSSEN, R. D. & LUDFORD, G. S. S. (1983a), The response to ambient pressure of fuel drop combustion with surface equilibrium, *International Journal of Engineering Science* (in press).

JANSSEN, R. D. & LUDFORD, G. S. S. (1983b), Burning rate response of a methanol drop to ambient air pressure, in *Nineteenth International Symposium on Combustion*, The Combustion Institute, Pittsburgh, pp. 999–1006.

MARGOLIS, S. B. & MATKOWSKY, B. J. (1983), Non-linear stability and bifurcation in the transition from laminar to turbulent flame propagation, *Combustion Science and Technology* (in press).

MARKSTEIN, G. H. (1964), *Nonsteady Flame Propagation*, AGARDograph No. 75, MacMillan, New York.

MATKOWSKY, B. J. & OLAGUNJU, D. O. (1982), Spinning waves in gaseous combustion, SIAM *Journal on Applied Mathematics*, 42, pp. 1138–1156.

MCCONNAUGHEY, H. V., LUDFORD, G. S. S., & SIVASHINSKY, G. I. (1983), A calculation of wrinkled flames, *Combustion Science and Technology*, 33, pp. 103–111.

MEYER, G. (1977), One-dimensional parabolic free boundary problems, SIAM *Review*, 19, pp. 17–34.

MIKOLAITIS, D. & BUCKMASTER, J. (1981), Flame stabilization in a rear stagnation point flow, *Combustion Science and Technology*, 27, pp. 55–68.

OLVER, F. W. J. (1964), Bessel functions of integer order, in *Handbook of Mathematical Functions* with *Formulas, Graphs and Mathematical Tables*, M. Abramowitz and I. A. Stegun, eds., National Bureau of Standards, Applied Mathematics Series 55. Superintendent of Documents, U.S. Government Printing Office, Washington, D.C.

PELCÉ, P. & CLAVIN, P. (1982), Influence of hydrodynamics and diffusion upon stability limits of laminar premixed flames, *Journal of Fluid Mechanics*, 124, pp. 219–237.

PETERS, N. (1982). *Private communication.*

ROGG, B. (1982), The effect of Lewis number greater than one on an unsteady propagating flame with one-step chemistry, in *Numerical Methods in Laminar Flame Propagation*, N. Peters and J. Warnatz, eds., Vieweg-Verlag, Braunschweig, pp. 38–47.

SIVASHINSKY, G. I. (1981), On spinning propagation of combustion waves, SIAM *Journal on Applied Mathematics*, 40, pp. 432–438.

SIVASHINSKY, G. I. (1983), Instabilities, pattern formation, and turbulence in flames, *Annual Review of Fluid Mechanics*, 15, pp. 179–199.

SIVASHINSKY, G. I., LAW, C. K., & JOULIN, G. (1982), On stability of premixed flames in stagnation-point flow, *Combustion Science and Technology*, 28, pp. 155–160.

SMITH, F. A. & PICKERING, S. F. (1929), Bunsen flames of unusual structure, *U.S. Bureau of Standards Journal of Research*, 3, pp. 65–74.